W9-DBJ-101

Measuring the Impact of Social Media on Business Profit & Success

This book is part of the Peter Lang Media and Communication list.
Every volume is peer reviewed and meets
the highest quality standards for content and production.

PETER LANG
New York • Bern • Frankfurt • Berlin
Brussels • Vienna • Oxford • Warsaw

CONG LI & DON STACKS

Measuring the Impact of Social Media on Business Profit & Success

A FORTUNE 500 PERSPECTIVE

HUMBER LIBRARIES LAKESHORE CAMPUS
3199 Lakeshore Blvd West
TORONTO, ON. M8V 1K8

PETER LANG
New York • Bern • Frankfurt • Berlin
Brussels • Vienna • Oxford • Warsaw

Library of Congress Cataloging-in-Publication Data

Li, Cong.
Measuring the impact of social media on business profit and success:
a Fortune 500 perspective / Cong Li, Don Stacks.
pages cm
Includes bibliographical references and indexes.
1. Social media—Economic aspects.
2. Business communication—Technological innovations.
3. Business enterprises—Information technology. I. Stacks, Don W. II. Title.
HM742.L4298 658.85—dc23 2015009558
ISBN 978-1-4331-2579-9 (hardcover)
ISBN 978-1-4331-2578-2 (paperback)
ISBN 978-1-4539-1590-5 (e-book)

Bibliographic information published by **Die Deutsche Nationalbibliothek**.
Die Deutsche Nationalbibliothek lists this publication in the "Deutsche Nationalbibliografie"; detailed bibliographic data are available on the Internet at http://dnb.d-nb.de/.

The paper in this book meets the guidelines for permanence and durability of the Committee on Production Guidelines for Book Longevity of the Council of Library Resources.

© 2015 Peter Lang Publishing, Inc., New York
29 Broadway, 18th floor, New York, NY 10006
www.peterlang.com

All rights reserved.
Reprint or reproduction, even partially, in all forms such as microfilm, xerography, microfiche, microcard, and offset strictly prohibited.

Printed in the United States of America

Contents

Preface

The idea of writing this book originated from a casual chat one afternoon about how social media are affecting everybody's life, both college professors and students alike. As an advertising professor and a public relations professor respectively, we have noticed something in common in our teaching: Students are showing a growing interest in knowing how to use social media for different purposes, such as a public relations campaign. Numerous industrial cases suggest that social media are becoming important for organizational communication, especially nowadays. However, people in different disciplines seem to disagree on how to measure the outcome of using social media. For example, advertising professionals may be more interested in the change of brand awareness whereas public relations practitioners may attach more importance to brand reputation. On the other hand, people in a management position tend to think financial outcomes such as sales and net incomes matter the most. Such a discrepancy poses a significant challenge to our teaching and students' learning.

What is the "right" way to teach students how to measure the outcome of an organization's activities on social media? Or, should we say, is there a "right" way to teach students how to measure the outcome of an organization's activities on social media? For example, an organization may have an impressive number of *"likes"* on its Facebook page and numerous *"followers"* on its Twitter account, but does that mean anything from a business performance perspective? In other words, is it worth an organization's effort to maintain an active presence on social media? The key question is: Is it reasonable to use social media metrics such as the number of Facebook *"likes"* and the number of Twitter *"followers"* to predict an organization's business "success" even though those metrics are nonfinancial indicators?

We consulted the literature for the answers to these questions. Many prior research studies discussed how an organization should utilize social media for various purposes. However, few studies provided strong empirical evidence to support how the outcome of using social media should be measured and why. In general, there is no conclusion or consensus on whether a "success" on social media such as generating Facebook *"likes"* can translate into a "success" in the real business world in the form of revenues and net incomes.

We then decided to conduct a pilot test to see whether we could dig further in this research direction and fill the theoretical gap in the literature. We chose to examine how the Fortune 500 companies use social media because they are the leaders in various industries. We searched for each Fortune 500 company's social media account on five different platforms including Facebook, Twitter, YouTube, Google+, and Pinterest. These social media platforms were selected because of their popularity. For those companies that had an account on a certain platform, we recorded several key product indicators (KPIs) from their account such as the number of Facebook *"likes,"* Twitter *"followers,"* and YouTube video *"views."* All these data were collected in March 2013. In addition, we recorded each public company's stock price and earnings per share on the second day after all the KPIs were collected. We also recorded each public company's net income in 2012. Based on these data, we conducted a series of correlation analysis. The results showed that a company's nonfinancial activities on Facebook, Twitter, YouTube, and Google+ were significantly associated with its Fortune 500 ranking, net income, and stock price. For example, when a company had more tweets on its Twitter account, it tends to have a higher net income and also ranks higher in the Fortune 500 list. A company's total number of subscribers on its YouTube channel is significantly and positively associated with its stock price. We did not find any significant relationship between a company's nonfinancial activities on Pinterest and its Fortune 500 ranking and other business performance indicators, possibly because the history of Pinterest is much shorter than the other four social media platforms.

The results of the pilot test gave us confidence in going further and conducting a more comprehensive study of how a "success" on social media might be associated with a "success" in the real business world. We decided to keep Facebook, Twitter, YouTube, and Google+ in this examination but dropped Pinterest due to the pilot test findings. To gain an in-depth understanding of how organizations use social media over the years, we decided to explore the Fortune 500 companies' social media activities (communication messages to the public) in a five-year time span from January 2009 to December 2013. Meanwhile, the business performance data of each public company during the same five-year period such as stock return, total revenue, net income, and earnings per share were also collected. These two types of data, both financial and nonfinancial, were then matched and statistically analyzed to see whether a company's social media activities were significantly associated with its business performance.

After we finalized this research design, we developed a book proposal and presented it to Peter Lang Publishing. We were awarded a contract after the proposal had been favorably reviewed by external reviewers. Since this research project aimed to cover 500 companies' nonfinancial activities on four social media platforms in five years, we anticipated to set up at least four databases, one social media platform each. Each database would have thousands of cases because we intended to randomly sample one communication message from each company in each month during the five-year time span (for example, if every company had an account on Facebook and posted at least one message each month from January 2009 to December 2013, we would have 500 × 60 = 30,000 cases in our Facebook database). We realized that this study was too ambitious to complete by just two researchers, and we needed help. Luckily enough, we had several talented doctoral students in our school, and we turned to them for assistance in the data collection. These students were amazingly kind and hard-working. They helped us tremendously in the raw data collection. The truth is we would never be able to finish writing this book on time if we did not have their help. We want to express our most sincere gratitude to them, including Jiangmeng Helen Liu, Michael North, Yi Grace Ji, Fan Yang, Zifei Fay Chen, Cheng Hong, Chun Zhou, and Qinghua Candy Yang for their incredibly warm-hearted and generous support.

Moreover, we would like to thank Dr. Tie Su in the Department of Finance of the School of Business Administration at the University of Miami for his valuable suggestions on how a public company's business performance should be measured, from both a finance and an accounting perspective. Many of our research questions were constructed based on his advice. We also want to thank the two anonymous reviewers of this book for their constructive feedback. Finally, we want to extend our thanks and appreciation to Mary Savigar, a senior acquisition editor at Peter Lang Publishing. Mary was very friendly and professional in communicating with us about how to prepare the book proposal and the final manuscript. This book would not come out so quickly without her efficient work.

Cong Li
Don W. Stacks
May 25, 2015

Chapter One

Introduction

Chapter Overview

In this opening chapter we will describe the emergence of social media and their impact on business operations. Specifically, we will discuss how social media have changed the way people communicate with and about business, and why measuring the influence of social media on business performance is necessary and important, but also difficult. Based on these discussions, we aim to answer the following questions:

- What are social media, and how are they evolved?
- What is the role of social media in the promotional mix within today's business environment?
- How have social media changed the way people communicate with and about business?
- Why is testing the impact of social media on business performance so difficult and what is our approach in this book?

Emergence of Social Media

It would be an understatement to note that the world has changed drastically over the last 3 decades. Countries have changed. Businesses have changed. Climate has changed. And the way we communicate has changed. Perhaps the most dramatic change, however, has been in communication—how people communicate among themselves and how they communicate with and about business especially. As noted in Figure 1.1, the communication changes have been particularly affected by technology: First, the printing press; second, the radio; third, the motion picture; fourth, the television; fifth, the computer, and finally, social media as an extension of the computer.

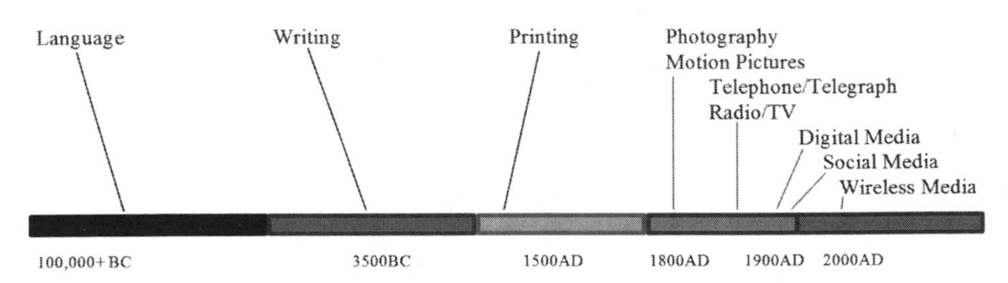

Figure 1.1. Historical timeline of mass communication.

It is this, the social media, which is the focus of this book. In a well-cited article (Kaplan & Haenlein, 2010), social media was broadly defined as a group of online applications that allow "creation and exchange of user generated content" (p. 61). Based on this definition, there exist at least six types of social media, including social networking sites such as Facebook, content communities such as YouTube, collaborative projects such as Wikipedia, blogging services such as WordPress, virtual social worlds such as Second Life, and virtual game worlds such as World of Warcraft. Among these six types of social media, social networking sites are probably the most visible and most frequently used by the public (Boyd & Ellison, 2007). As seen in Table 1.1, more than a dozen social media were ranked among the top 50 most frequently visited websites in the world according to a website-traffic ranking by Alexa (2014).

Table 1.1. Most Frequently Visited Social Media Ranked by Alexa

Facebook	Social networking site
YouTube	Video-sharing site
Wikipedia	Online encyclopedia
Twitter	Microblog
QQ	Instant messaging service in China
LinkedIn	Social networking site for professional connections
Weibo	Microblog in China
Blogger	Blog
VK	Social networking site in Russia
WordPress	Blog
Reddit	News-sharing site
Pinterest	Visual-based social networking site
Tumblr	Microblog
Instagram	Photo-sharing site
Imgur	Photo-sharing site

The most important function of social media is to help people connect with each other and share contents (Boyd & Ellison, 2007; Raacke & Bonds-Raacke, 2008). There are many studies that talk about how such connections and sharing affect people's lives and behavior. A recent example can be found in an article published in *Nature* (Bond et al., 2012), which showed that messages posted on Facebook influenced people's political self-expression, information-seeking, and real-world voting. As social media become more popular, business enterprises also start to pay attention to these media platforms. There have been many discussions on how businesses should use social media to communicate with their stakeholders and how managers should leverage social media for different business purposes (e.g., Berthon, Pitt, Plangger, & Shapiro, 2012; Chaney, 2009; Christ, 2005; Dutta, 2010; Edosomwan, Prakasan, Kouame, Watson, & Seymour, 2011; Lin & Lu, 2011; Tuten & Solomon, 2013; Waters, Burnett, Lamm, & Lucas, 2009). However, to date, there have been no long-term *analyses* of social media impact on financial outcomes; instead, most research discusses *uses* of social media, and leaves it to others (such as us) to figure out how social media impact on the business. As an *MIT Slogan Management Review* article (Hoffman & Fodor, 2010) proposed, how to measure the return on investment (ROI) of an organization's social media activities is the key question that remains to be answered.

That is exactly the objective of this book. We believe we are the first to *demonstrate* the impact of using social media on business performance outcomes, based on a large-scale and longitudinal investigation. In particular, we asked this question: Does the use of social media by the Fortune 500 companies affect their business "success" or "failure"? Stated differently, do businesses who engage with their stakeholders (e.g., customers, employees) through various social media platforms (e.g., Facebook, Twitter, YouTube, and Google+) demonstrate financial and reputational strength?

The Role of Social Media in the Promotional Mix

One of the defining problems faced by today's businesses is determining just what the social media are and then what role they play in the traditional promotional mix of advertising, marketing, and public relations/corporate communication. The social media, though still a recent tool for getting information out to various target audiences, have evolved from a traditional mass media tool to one that more closely approximates an extended interpersonal tool. What we mean by "extended interpersonal" is that social media now approximate what was once considered to be interpersonal communication: It is one-to-one (with many potential "interactants"), it is immediate (or close to immediate), and it is evolving to a face-to-face type of interaction.

Historically, social media began with the advent of the Internet way back in the late 1970s as a way to connect different "nodes" together for sharing information. The introduction of the personal computer on the Internet through portal connections via modems and software such as Red Ryder in the early 1980s allowed messages to be sent across the Internet and the sender to follow them as they passed through different university portals to their intended recipients. This grew into what we now know as the World Wide Web, with multiple portals that allowed for faster and faster transmission of messages, and as bandwidth and storage space increased, the capability to reach more people with more developed kinds of communication tools offered new and exciting ways for businesses to engage their stake- and stockholders in conversations, get their impressions of products, and generally provide another persuasive way to increase profits. What we used then—e-mails or electronic letters or memos, primarily—have become a multimedia tool reaching millions of people in places never considered before due to language and access problems.

In general, what began as information sharing became promotional messages with a strong persuasive element. The change from an electronic letter or memo to one or more recipients with the option to copy or blind copy others to today's open-ended communication tool that provides immediate reaction (e.g., "liking" something) has drastically changed how companies approach their business objectives. But this still has not answered the question, "What is or are the social media, from a business communication perspective?" A simple answer might be, "platforms that allow individuals and organizations to communicate and promote ideas, brands, products, and organizations to others." This definition fits the needs of most, but given social media's movement towards a more interpersonal channel, it takes on a symmetrical nature; in this channel, customers and businesses engage in discussions, make recommendations, and influence others who might not necessarily be part of that discussion. Thus, from a promotional mix approach, communication becomes more "personal."

This personalization of business communication changes how traditional advertising, marketing, and public relations are practiced. Traditional approaches to the mix, sometimes called integrated marketing communication (IMC), dictated that the promotional function most appropriate for a problem takes the lead and devises a business communication strategy. As the unit that is actually a part of the business, marketing almost always took the lead (and received the largest portion of the money and team budget). In the "old days" the strategy was basically that marketing set the strategy, advertising (paid placement of messages) sent messages to targeted audiences through traditional media (e.g., print, broadcast) in the form of advertisements, and public relations was often employed as a way to get the advertised messages out through "word of mouth." The social media have changed this linear process into a continual feedback loop whereby sender and receiver are no longer distanced by time and space; now they are engaged in mutually beneficial relationships.

The Changing Business Environment

The social media have basically revolutionized how business has to operate. No longer is business done as usual, following *caveat emptor* (let the buyer beware). Instant reaction to business is now the way things are done. A product that does not work—instant criticism. A brand that is not living up to its values—instant criticism and often elaboration. A company that does not offer the services customers expect—instant complaints. Even how businesses define their audiences has changed due to social media. Customers, employees, regulators, and so forth are now important stakeholders with whom the business must engage in a continued dialogue through a variety of social media platforms. Additionally, the stakeholder who back in the day had little or no say about a business's products, how the business was run, and/or how its brand was promoted now has a venue to comment on both positive and negative experiences and have those messages read by people worldwide who tag their own experiences onto the original messages.

When looking at today's interconnected world it is clear that social media have had a profound impact on stakeholder decision making related to business. Employees can support or complain about a company's decision on just about anything it does.[1] Customers now have direct lines to let others know how they feel about a product or brand. Regulators can monitor social media looking for potential problems. Basically, social media now impact on five important business outcomes: reputation, relationship, trust, credibility, and confidence (see Figure 1.2). These in turn impact on a business's bottom line (see Figure 1.3).

What makes these outcomes so important is that they are all found in the eye of the stakeholder (and stockholder, too). They are formed through the communications directed at stakeholders that yield expectations of that business. The business and communication literatures are stocked with articles that point to the importance of establishing expectations and then maintaining or increasing those expectations. As in the financial side of things, a negative review of a product or a less-than-spectacular earnings report leads to negative expectations. Negative expectations lead to decreased stock prices, plummeting sales, and often, leadership changes. We will look more closely at these five outcomes in Chapter 2.

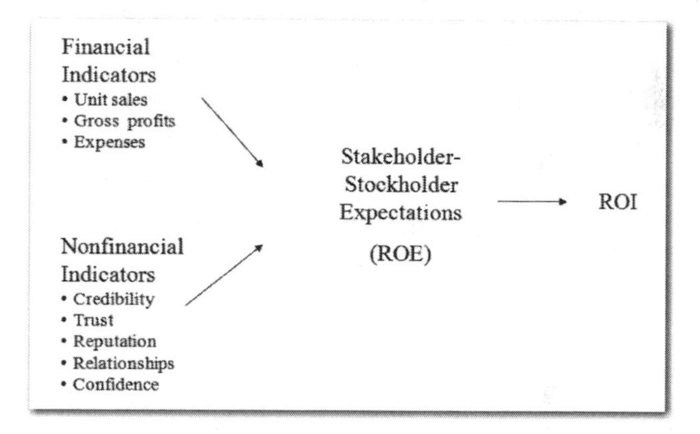

Figure 1.2. Relationship of financial and nonfinancial indicators to ROI. Reprinted from *Primer of Public Relations Research* (2nd ed.), by D. W. Stacks, 2011, New York: Guilford. Copyright 2011 by Guilford. Reprinted with permission.

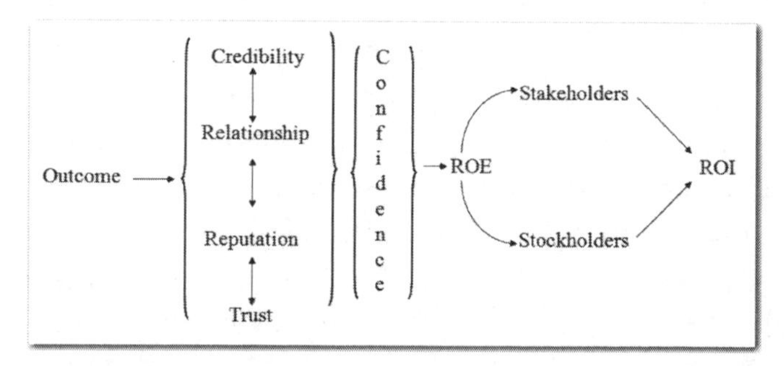

Figure 1.3. Basic model of nonfinancial variables. Reprinted from *Primer of Public Relations Research* (2nd ed.), by D. W. Stacks, 2011, New York: Guilford. Copyright 2011 by Guilford. Reprinted with permission.

Thus far we have stayed away from defining specific social media platforms that have become important to business and stakeholder expectations. Although there are many platforms that have come and gone over the past 25 years, the four that have endeared themselves to target audiences and hence sparked business interest are easily identified: Facebook, Twitter, YouTube, and Google+. Each allows both businesses and stakeholders to engage in relationships, each allows them to set strategies for uses, and each is thought to have direct impact on business outcomes. However, no study to date has actually tracked across businesses the impact these platforms have on financial outcomes. Why? First, the volume of social media use is tremendous. It takes time to gather data and analyze and evaluate them. Second, what it is that research should look at in terms of social media outcomes is still not quite understood, especially, what particular outcomes mean or are worth from a measurement perspective. And third, to whom should the outcomes be applied?[2]

Our Approach in This Book

As described earlier, social media are profoundly affecting today's business operations, but the long-term effects of using social media to engage stakeholders on an organization's business performance have not been systematically examined. Our goal in writing this book is to shed new light in this research direction, and we hope to generate some solid conclusions based on strong empirical evidence. We undertook a review of recent advertising, marketing, and public relations/corporate communication literature to understand the effects of strategic communication use of social media on business outcomes (e.g., Barwise & Meehan, 2010; Hansson, Wrangmo, & Søilen, 2013; Lillqvist & Louhiala-Salminen, 2014; Mamic & Almaraz, 2013; Nelson-Field, Riebe, & Sharp, 2012; Paek, Hove, Jung, & Cole, 2013; Rybalko & Seltzer, 2010; Saxton & Waters, 2014; Smith, Fischer, & Chen, 2012; Swani, Milne, & Brown, 2013; Toubia & Stephen, 2013; Wallace, Buil, de Chernatony, & Hogan, 2014).[3] Based on this review, we decided to test the association between the Fortune 500 companies' social media activities and their business outcomes. We focus on the Fortune 500 companies because they are the leaders in various industries, and they tend to be admired for the way they operate their businesses (supposedly including how they use social media to engage stakeholders). The fundamental research idea, then, is to examine the relationships between the nonfinancial indicators associated with the Fortune 500 companies' social media activities and the financial indicators associated with their business outcomes.

As noted, the nonfinancial indicators we will be looking for in the social media include reputation (e.g., likes, favorites), relationship (e.g., replies, comments), trust (e.g., retweets, shares), credibility (e.g., followers, subscribers), and confidence (e.g., reply valence, comment valence).[4] These nonfinancial indicators serve as variables that set stakeholder expectations. Of particular interest is the impact of what has been called the "third-party endorsement" effect, whereby public relations strategy employs a target audience's influencers to modify or reinforce target audience behavior. Third-party endorsers include close friends, professionals not associated with the company or brand, particular traditional and social media writers, and even, at times, a company's employees themselves.

In this book we define business outcomes quite broadly. As will be further discussed in Chapters 2 and 3, we used financial indicators such as a company's monthly stock return, quarterly total revenue, net income (or loss), earnings per share, and so on to assess its business "success" or "failure." To obtain these data, we used a variety of sources. As seen in Table 1.2, the primary objective of this book is to test and demonstrate the relationships between the two groups of indicators, nonfinancial and financial.

Table 1.2. Primary Objective of the Data Analyses

A company's social media activities (Nonfinancial indicators)	Relationships of the two	A company's business outcomes (Financial indicators)
Reputation (e.g., likes, favorites)		Monthly stock return
Relationship (e.g., replies, comments)		Quarterly total revenue
Trust (e.g., retweets, shares)		Quarterly net income (or loss)
Credibility (e.g., followers, subscribers)	?	Quarterly earnings per share
		Quarterly profit margin
Confidence (e.g., reply valence, comment valence)		Quarterly return on assets
		Quarterly return on equity

To determine what social media platforms to include in these analyses, we conducted a pilot test before we finalized the research design for this book. The pilot test was a one-shot study aiming to provide some useful information on how the Fortune 500 companies use social media in general, and whether using social media may relate to their business outcomes. We included five social media platforms in the pilot test based on their popularity among the public: Facebook, Twitter, YouTube, Google+, and Pinterest. In particular, we searched for each Fortune 500 company's accounts on those platforms to see how well they were accepted and adopted for corporate communication. We collected a handful of nonfinancial indicators from the companies' accounts on those platforms, such as the numbers of "likes" on Facebook, "tweets" on Twitter, video "views" on YouTube, "plusses" on Google+, and "pins" on Pinterest. We also recorded a few financial indicators for each public company, including stock price, earnings per share, and net income (or loss).[5] Based on analyses of these data, we drew two conclusions. First, although the pilot test was limited in its research scope, it did show that a company's nonfinancial activities on Facebook, Twitter, YouTube, and Google+ were somewhat significantly associated with its Fortune 500 ranking, stock price, and net income (or loss).[6] Second, the adoption rate of Facebook, Twitter, YouTube, and Google+ among the Fortune 500 companies was quite high (see Table 1.3). Overall, to a certain extent all industries/business sectors deem social media to be important tools to communicate with their stakeholders.[7]

Based on the pilot test results, we decided to conduct another study in the same vein but on a much larger scale, which eventually became our research plan for this book. In this new research, we kept the Fortune 500 companies as our sample and examined their usage of Facebook, Twitter, YouTube, and Google+ in a 5-year time span from 2009 to 2013 (we dropped Pinterest due to the pilot test findings). Such a longitudinal design provided us more flexibility in data analyses such as testing trends across different years. The specific research methodology will be explained in more detail in Chapter 3.

Table 1.3. Business Sectors with Social Media Accounts

Business sectors	Total # of companies	Sample companies	# on Facebook	# on Twitter	# on YouTube	# on Google+
Energy	39	Exxon Mobil, Chevron	19	22	20	19
Materials	35	Dow Chemical, DuPont	25	25	25	22
Industrials	71	GE, Boeing	62	58	55	54
Consumer discretionary	90	GM, Home Depot	81	80	70	71
Consumer staples	46	Walmart, CVS	38	41	36	31
Health care	48	McKesson, UnitedHealth Group	30	42	37	31
Financials	51	Wells Fargo, AIG	40	43	41	39
Information technology	48	Apple, Cisco	47	47	48	45
Telecommunication services	10	AT&T, Verizon	10	8	10	10
Utilities	26	Exelon, Duke Energy	24	24	19	16

Summary

There is no denying that social media have become an integral part of business today and that its impact on business outcomes—financial, of course, but also nonfinancial—has a direct effect. In Chapters 2 and 3 we will do a more in-depth analysis of social media strategy and explain how our research was actually conducted. Chapters 4 to 7 present the results for each social media platform, and Chapter 8 looks at all four platforms as a whole and provides more theory-driven data analyses. Finally, Chapters 9 and 10 discuss the findings, their implications for both business and the related promotional functions of advertising, marketing, and public relations, and the future directions for social media measurement.

Chapter Two

Research Background

Chapter Overview

In this chapter we will present the theoretical framework of our research and discuss how social media have changed the practices of advertising, marketing, and public relations/corporate communication nowadays. Several key concepts will be introduced, such as third-party endorsement, owned media, paid media, earned media, outputs, outtakes, and outcomes. Specifically, we intend to address the following questions in this chapter:

- What is third-party endorsement?
- What are the differences between owned media, paid media, and earned media?
- How are social media related to the effects of third-party endorsement?
- What is the communication lifecycle, and how have social media affected it?
- What are the differences between outputs, outtakes, and outcomes?
- What are the traditional business metrics?
- What are the social media metrics?

Third-Party Endorsement

Traditional promotional communication campaigns focus on the placement of messages in front of target audiences. In general, marketing focuses on the brand and its reputation. Advertising creates advertisements that are placed in specified newspapers, magazines, television shows, and so forth. These are paid placements, and estimates of "value" can be made by how many people have an opportunity to see (OTS). The estimate is based on how many newspapers or magazines are sold (circulation) or how many viewers or listeners are logged onto the television or radio. Public relations traditionally focuses on having others carry the message to target audiences and serve as sources of influence that are more credible since they are not identified as "paid" or "owned" messages, but instead are seen as "earned media."

The distinction between the three types of media is critical when examining how social media have changed the promotional landscape in terms of what each does and how each is measured. According to Corcoran (2009), *owned media* is associated with the company, brand, or product, and is typically seen as "marketing" since its function is brand control. It is created and controlled by that company and serves to build relationships among information seekers. Owned media's downside is that it is what the company wants people to see, thus it may not be trusted as a neutral source of information. *Paid media* is most often associated with advertising. While it often has direct emotional appeal, paid media has poor credibility (*caveat emptor*) and is traditionally a source-to-receiver, one-way message channel. *Earned media* occurs when two-way communication channels are made available to information seekers. Traditionally, public relations has employed the earned media function in promotional campaigns, encouraging influencers to pass on the key messages of the campaign for the company, brand, or product. It is the most credible of the three types of media, but lacks control of the message once that message has been released to the targeted audiences.

It is the role of influencers that is our concern. Influencers have been the target of public relations campaigns since the early twentieth century. Legendary public relations professional Edward Bernays first introduced the concept of "third-party endorsement" in 1913 with a pioneering strategy to create an intermediary party to endorse the client, in this case, the play *Damaged Goods* (Cutlip, 1994). Third-party endorsement can take many forms. In early public relations endorsers were basically front organizations that supported and released the key messages that the promotional campaign wanted target audiences to hear. Later, third-party endorsement focused on specific individuals who were associated with the campaign—often celebrities who took up a cause and endorsed its messaging. Advertising also used endorsement and still does, but the endorsers are typically high-profile individuals who have been paid for their endorsements (or they have not been paid, but the audience may *perceive* that they have been paid, and thus they are less credible).

Traditional, non–social media third-party endorsement seeks to influence target audiences through earned media by (1) identifying who the key influencers are, (2) establishing a relationship between the company's public relations professionals and influencers, (3) providing these influencers with materials that they can use in their messages, and (4) tracking those influencers' messages to key constituents. The key here is that the messages do not come from the company whose brand or product is being talked about, but instead from the key influencer. A mass media theory called *agenda setting* (McCombs & Shaw, 1972) suggests that the amount of time the media spend on a subject can be correlated to its perceived importance. A key influencer spending time talking or writing about something will increase its importance to his or her audience because of the perceived credibility the key influencer brings to the situation (Davidson, 1983; Neuwirth & Frederick, 2002). Key influencers often take the roles of news reporter, industry analyst, academic expert, and so forth.

The major disadvantages of third-party endorsements take two forms. First, the flow of messaging is one-way; the messages are not easily responded to in real time. Second, once the key influencer's messages have been provided to his or her audience, control is lost over how those exposed to the messages are (1) responding to them and (2) extending their reach through individual-to-individual transmission or word of mouth (WOM). This model of influence suggests that individual audience members become informal (or formal, depending on the relationships) influencers who have taken the time to communicate to others or have provided requested evaluations about a company, brand, or product. Traditional media approaches do not allow for this two-way form of communication to occur.

What Do Social Media Add to the Equation?

Think of social media as a mediator of expectations that allows stakeholders to add voice to the communication about a company, brand, or product.[1] Think further that social media platforms provide both companies and stakeholders a real-time, relational database of nonfinancial data that may or may not impact on the company. Thus, social media provide a way for stakeholders to compare notes, to provide information on companies and products in a way never before imagined. Social media platforms provide the channel by which to praise or damn a company or product. Social media allow stakeholders to get opinions from sources that are not directly tied to the companies or products (DiStaso & McCorkindale, 2013). Social media provide us with ways to look at what Dr. David Michaelson (Michaelson & Stacks, 2014) has called the "communication lifecycle" in a totally new way (see Figure 2.1).

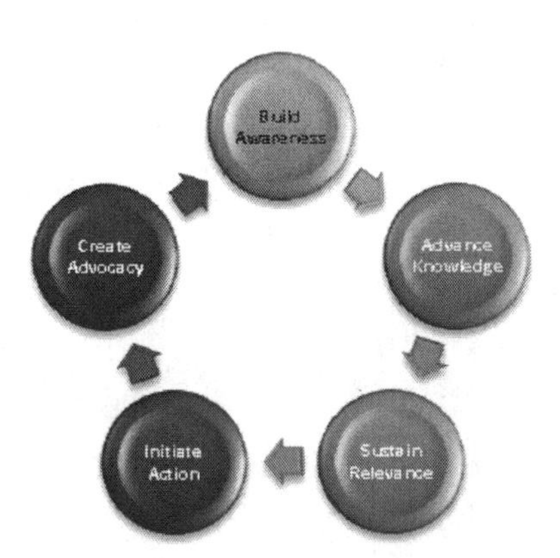

Figure 2.1. The communication lifecycle. Reprinted from *A Professional and Practitioners Guide to Public Relations Research, Measurement, and Evaluation* (2nd ed.), by D. Michaelson and D. W. Stacks, 2014, New York: Business Expert Press. Copyright by D. Michaelson. Reprinted with permission.

All promotional communications have a lifecycle during which stakeholders first engage with a company or product. For a traditional communication campaign, the goal is to get people to advocate for the company or product through word of mouth or interpersonal contact with their friends. This outcome is often referred to as the "third-party endorsement" in public relations campaign strategy (Michaelson & Stacks, 2014). From the traditional promotional perspective, the company's advertising, marketing, and public relations researchers attempt to define where the company target audience is in the communication lifecycle. If the object of communication is new or relatively unknown, then the communication strategy will be to build awareness. If the target audience is aware, then the strategy is to build knowledge by adding to the audience's understanding of the product or brand or company. If it is aware and knowledgeable, the strategy is to sustain relevance for the product or company—to keep it in the stakeholder's mind's eye. When all three parts of the lifecycle have been accounted for, the strategy is to initiate action, to get the stakeholder to actually begin taking actions, such as communicating with others and raising interest in his or her own sphere of influence. Finally, the last stage of the lifecycle is when the stakeholder actually advocates for the company or product or brand.

From a social media perspective, how does the lifecycle operate? While traditional promotional campaigns seek to target specific audiences at various stages of the lifecycle, social media provide nontargeted audiences to create their own awareness of a company, brand, or product. In addition, social media build awareness in both targeted audiences (aware and unaware), by reaching them via paid campaigns, and nontargeted or unintended audiences, via electronic word of mouth (eWOM), although our focus is on the latter.[2] In essence, the campaign strategy is to establish two-way communication channels whereby the company can in real time identify influencers both formal and informal and track how the communication is being diffused within and across platforms.

A simple example might be a new product being offered to a targeted set of influencers (these are our third-party endorsers, or from a diffusion of innovations [Rogers, 2003] perspective, those who are innovators or early adopters) who communicate through social media about their experiences with the product. This is basically what a traditional newspaper or magazine review of a product does. However, in the case of social media, the unintended target can become aware by following the influencer personally (or even through one of his or her friends who follows, shares, or retweets the influencer's social media comments), create a personal knowledge base, determine whether it is relevant, initiate action (commenting on the comment within his or her own social media community and within and between platforms), and advocate some form of action.

Clearly, social media expand the traditional one-way communication to the two-way relational model advocated by public relations professionals (e.g., Grunig, 1984; Grunig, Grunig & Dozier, 2002). Social media, however, also offer the advertising and marketing professional tactics to impact on the communication lifecycle in the traditional way (i.e., owned media). They may drive awareness by optimizing search engine strategy, purchasing "banners" or social media advertisements on a platform, and so forth. Note that these strategies do not call for interaction, just awareness or sustainability of the company, brand, or product. Advertising a specific social media post may generate a greater reach, and larger numbers of likes and comments, thereby allowing more interaction with the target audience. The two-way model, however, can be used to sustain or enlarge the advertising and marketing strategy by creating an informed social media policy of its own (i.e., earned media).

Companies that have clear social media policies are generally perceived as more attractive than companies that do not (Cho, Park, & Ordonez, 2013). In particular, companies that follow their brand on social media platforms can "interact" with target audiences and "react" to unintended audiences who may have concerns or problems with a product or brand. This allows them to follow in real time positive, neutral, or negative messaging occurring on social media. For example, one of the authors was caught in the Heathrow snowstorm of 2010. For 3 days he slept on the floor of London's Heathrow 5 terminal. In a tweet he suggested that "anyone flying British Airways always carry a sleeping bag" because British Airlines did nothing for passengers who were stranded. A day later, British Airlines passed out foam mats to passengers to lie on—mats but no sheets or blankets. Thus, the company used the tweet to correct a situation that could have gone badly and really impacted on its credibility, trust, and confidence in its stakeholder audience, and would have required it to rebuild relationships and reputation among the flying public.

How important are social media to today's business community? McKinsey & Company found that 56% of executives say that digital engagement is at least a top-ten company priority (2013). Three out of four companies have social media teams (Altimeter, 2013). Almost 40% of social media users have purchased after perusing social media platforms (VisionCritical, 2013). The Institute for Public Relations (www.instituteforpr.org) has a body of social media findings that clearly indicate that social media impact on a company's bottom line.

Earned Media as a Function of the Social Media

Earned media has taken a new and more important function in promotional campaigns. Although both marketing and public relations work in the social media, roles have been reversed. In the 1980s "integrated marketing media" was conceived (Duncan & Caywood, 1996; Schultz & Schultz, 1998) as a model working relationship whereby the promotional function (primarily marketing and public relations, as advertising was seen as falling under the marketing function) that should best address the promotional problem would take the lead. However, the reality was that marketing very often resided in a company's senior management team (led by the Chief Marketing Officer, or CMO) with other *business* management leaders (e.g., the Chief Human Resource Officer, or CHRO; the Chief Financial Officer, or CFO) who developed strategy, including communication strategy. At that time there were few Chief Communication Officers (CCOs) serving on company management boards, and most communication teams reported to the CMO or the CHRO or CFO.[3] With the advent of social media, the focus has turned to what Dr. James Grunig (1984; Grunig & Grunig, 1992; Grunig, Grunig, & Dozier, 2002) labeled "two-way symmetrical communication," or communication that more closely resembled a conversation. Social media have taken this concept to a real-time engagement, and one that business has had to apply as customers, employees, stockholders, and others now have platforms on which to comment on companies, brands, and products. Such commentary often requires immediate reaction to reduce negative images or manage crises (Capozzi & Rucci, 2013).

There are at least eight social media platforms that people currently use in making contact with companies and each other:

- Facebook
- Twitter
- YouTube (owned by Google)
- Instagram (owned by Facebook)
- Pinterest
- Tumbler (owned by Yahoo!)
- Vine (owned by Twitter)
- Google+ (owned by Google)

In reality there are hundreds, if not thousands, of social media platforms existing on the Internet today. The platforms listed above represent those with strong impacts on business today. They do not differentiate between nationality and/or culture, and have between 40 million and more than 1 billion users. In this book we are analyzing four of them: Facebook, Twitter, YouTube, and Google+. Each platform has generally accepted key product indicators (KPIs) that allow for estimations of impact. It should be noted that these social media platforms provide two different types of posts. Companies may employ *paid* posts that are targeted to specific demographic audiences, and they can hire social media research companies to provide specific KPIs. *Organic* posts are posted but not promoted as such, and may not get the most clicks, likes, comments, and so forth. Because we have no way of distinguishing paid posts from organic posts, we are classifying all posts as organic.

Measurement Conundrum

Public relations measurement beyond simple counts of production units (media outputs, mainly) and its impact on a client's or company's fiscal outcomes did not come about until the early 1990s. Some public relations firms and a few companies did attempt to gauge through measurement some sort of impact, but it was neither systematic nor standardized. In the mid-1990s the Institute for Public Relations (IPR) created its first commission to study this problem, the IPR Measurement Commission, as it is now known. Under its auspices, a number of white papers and publications were produced that examined (1) what public relations should measure, (2) how that measurement should be accomplished, and (3) what standards public relations measurement should be compared to in order to assess reliability and validity of those measures.

The Commission's objective was to take the conundrum out of measurement, to set up best practices (early in its existence), and to produce standards against which measurement could be assessed. Early leaders, under the direction of Jack Felton at the Institute, included Katie D. Paine, Mark Weiner, Patrick Jackson, Donald K. Wright, David Michaelson, James Grunig, Bruce Jeffries-Fox, and others who represented the research and measurement leaders of that time.[4] Their carefully considered works are too numerous to list, but a look at the publications found on the Institute for Public Relations' website provides insight into the way the measurement conundrum has been attacked and interesting and informed reading across a spectrum of public relations and marketing areas.

Perhaps the best known of the Commission's early work was the *Dictionary of Public Relations Measurement and Research*, edited initially by Don W. Stacks and in its current edition by Stacks and Shannon A. Bowen. This publication started the standardization process by defining terms used in research and measurement, and it has been translated into multiple languages. The Commission was early in taking leadership of how to measure social media impact. Importantly, Commission members began to focus on more than simple outputs, and they helped open the discussion to impact of social media on client/company nonfinancial and financial outcomes. At the forefront of this movement are Katie D. Paine, Mark Weiner, and David Geddes. Their work yielded new working groups that have taken on social media measurement, including standardizing social media measurement across different social media platforms. Additionally, their work, combined with

Stacks's theory positing that ROE → ROI, has produced a way of establishing public relations financial impact.[5]

Not much is known at this time about social media's impact on financial outcomes. It makes sense, however, that social media use should and would affect a company's financial outcomes. As channels for influence, social media allow an entire mix of audiences to participate in promoting business nonfinancial outcomes such as confidence, credibility, relationship, reputation, and trust. These outcomes, in turn, should have a demonstrated impact on financial outcomes as reported by publicly traded companies.

Monitoring of social media and engaging stakeholders in real-time communications should correlate with financial outcomes for companies that are engaged in social media or have social media policy and strategy as part of their total promotional planning. The problem, however, is confounded by the metrics used when collecting nonfinancial social media data, and confusion about those metrics produces what we call "the measurement conundrum."

There are a number of problems when it comes to social media measurement, but before we examine them, let's examine the concept of measurement in more detail. First, what is measurement? According Michaelson and Stacks (2014), measurement is the assigning of numbers to categories. As such, measurement occurs in two basic classes, each broken into two levels. The first simply puts observations (anything that you are interested in) into categories. The most basic, *nominal*, simply creates data that are different from each other as subcategories. An example would be the uses of Facebook, Twitter, YouTube, and Google+. At this level the data are simply considered different and could be coded as 1, 2, 3, or 4, or could be coded as "yes (1), no (2)." At a higher level of measurement, still looking at categories, is *ordinal* measurement. The Fortune 500 companies list examined in this book is an example of ordinal measurement, which assigns a rank value to each company. Rank order can be largest to smallest, highest to lowest, and so forth. Categorical measurement is evaluated in terms of actual number per category, percent of total, and proportion of total.

In contrast, the second measurement class, continuous measurement, assumes that observations occur on a line from start to finish.

Interval measurement assumes that the difference between units being measured is one unit apart. Thus, we are looking at measurement over the entire period as discrete units. An example might be the number of hours worked, when measured by the hour; if hours worked was measured by quarter hours, it would be an interval of 15-minute units. At the second level, ratio measurement, the data are extremely precise and can pass though a zero unit into the negative. This type of measurement is usually not found in the social sciences, but in this book, where financial data are being observed for each company, very precise measurement is possible—and even in the negative range. An example of ratio measurement might be a bank account, which can range from 0 (no money) to whatever amount is in the account to negative money, where an overdraft has occurred and the account is now in the negative (e.g., –$100.00 if the bank has covered the overdraft).

We will expand this discussion in Chapter 3, but first we need to consider what promotional communication actually measures. Basically, there are three different kinds of measures: outputs, outtakes, and outcomes.[6] *Outputs* are what are produced—advertisements, media releases, blogs, tweets, and so forth. *Outtakes* are intermediary measures that track progress in a promotional campaign. They focus on how well influencers and others are picking up and responding to messages. In social media these would take the form of clicks, likes, retweets, shares, views, and so forth. *Outcomes* are indicators of success of the promotional campaign and can be correlated with other indicators of success such as reduced absenteeism, product quality, immediate response to critical commentary, and how these relate to the financial status of a company. Next we will describe what metrics are available to measure outputs, outtakes, or outcomes in our research context (specifically, social media), either financial or nonfinancial metrics.

Traditional Business Metrics

Measuring business outcomes is fairly straightforward. The variables and data are what Stacks (2011) has called "hard." That is, they represent observations that can be seen: units produced, units sold, profits, losses, stock price, and so forth. Additionally, there are a number of generally accepted measures—or standards (see Michaelson & Stacks, 2014)—that can be used to evaluate businesses against each other. The measures employed in this study serve as the dependent variables against the social media platform measurements for association tests. Specifically, we looked at seven financial variables obtainable for publicly traded companies:[7]

- Monthly stock return
- Quarterly total revenue
- Quarterly net income (or loss)
- Quarterly profit margin
- Quarterly earnings per share
- Quarterly return on assets
- Quarterly return on equity

The data in our research were obtained from two databases (CRSP and Compustat) and represent a 5-year time span, thus taking into account the variability of the market and providing a solid set of data from which to draw conclusions. Table 2.1 presents a sampling of the financial data for three companies on the 2013 Fortune 500 list.

Table 2.1. A Sampling of Three Companies' Financial Data in 2013

Company	Year & quarter	Total revenue (in millions)	Net income (in millions)	Earnings per share
Exxon Mobil	2013 Q1	$95,886	$9,500	$9.82
	2013 Q2	$95,498	$6,860	$7.96
	2013 Q3	$100,508	$7,870	$7.66
	2013 Q4	$98,355	$8,350	$7.37
General Motors	2013 Q1	$36,884	$1,175	$2.91
	2013 Q2	$39,075	$1,414	$2.80
	2013 Q3	$38,983	$1,717	$2.36
	2013 Q4	$40,485	$1,040	$2.38
AT&T	2013 Q1	$31,356	$3,700	$1.28
	2013 Q2	$32,075	$3,822	$1.33
	2013 Q3	$32,158	$3,814	$1.41
	2013 Q4	$33,163	$6,913	$3.39

Social Media Metrics

While financial metrics and data are "hard," social media metrics represent a combination of hard and "soft" data. That is, there are key product indicators (KPIs) that could be considered hard, such as number of subscribers, total count of likes, and numbers of tweets and retweets. These data require a simple counting and represent continuous data that can be aggregated. Other data require some type of interpretation, such as comment *valence*.

While the metrics appear to be simple, to examine them from a nonfinancial perspective they need to be placed into promotional communication variables. As noted earlier, we are interested in how five nonfinancial variables impact on a company's financial success. Creating metrics for company perceived credibility, reputation, trust, relationship, and confidence from the data present on the various platforms requires operationalizing each of the nonfinancial variables in terms of the platform content. Operationalizing allowed us to inspect what was posted and create a series of KPIs that we believe reflect each nonfinancial variable.[8] More of this will be discussed in Chapter 3, but for now, the following social media metrics were observed:

- Facebook
    - General (for each company account): total number of post likes
    - Specific (for each post): numbers of post likes, shares, and comments, and comment valence
- Twitter
    - General (for each company account): total numbers of tweets, followings, and followers
    - Specific (for each tweet): numbers of tweet favorites, retweets, and replies, and reply valence
- YouTube
    - General (for each company account): total numbers of subscribers, views, and videos
    - Specific (for each video): numbers of video views, likes, dislikes, shares, and comments, video length, and comment valence

- Google+
  - General (for each company account): total numbers of followers, views, and people in the circle
  - Specific (for each post): numbers of post plusses, shares, and comments, and comment valence

Summary

We seek to identify and understand what defines success in terms of social media communication on business financial outcomes. To date, no literature exists that we can find that has addressed this specific problem or has utilized a set of data covering a 5-year timeline. This study allows us to examine the relationship between financial success and two-way interactive social media by the Fortune 500 companies in terms of company social media platform outputs, reaction to those outputs by stakeholders actively participating in the discussions (outtakes), and how the number and valence of those discussions interact with seven financial variables commonly used to follow success or failure in the marketplace. In the next chapter we will describe our study methodology in detail.

Chapter Three

Research Methodology

Chapter Overview

This chapter will present the methodology employed in our research. In particular, we will describe how we acquired the raw data, both financial and nonfinancial, how we evaluated the data, and also how we statistically analyzed the data to answer our research questions. The sampling procedure used in the research will also be discussed. In this chapter we aim to address the following methodology-related questions for the readers:

- How were the enterprises chosen for our research, and what industries did they represent?
- What social media platforms were chosen, and why?
- What methods were employed to collect the raw data, both financial and nonfinancial?
- How were the data coded in the database?
- What statistical techniques were used in the data analyses?

Key Research Question

One of the biggest problems today's promotional communication professionals have is demonstrating the impact on financial outcomes. As noted in Chapter 1, the ability to correlate actual communication impact on financial outcomes has always been problematic, even when considering traditional advertising and marketing. For the public relations function this correlation is even more problematic. As noted, the reasons the public relations function focuses on nonfinancial data are: company or brand credibility, stake- and stockholder confidence in that brand or company, relationship to company or brand, brand and company reputation, and trust in the company and brand. Although these nonfinancial variables can be found throughout the literature to be influencing stake- and stockholder expectations and, ultimately, decisions, there is a dearth of research that has looked at actual impacts of the newest communication channels associated with social media.

In this book we focus on four popular social media platforms: Facebook, Twitter, YouTube, and Google+. We look at a time series of data collected over a 5-year period from 2009 to 2013 and examine what these platforms have done for company return on investment, as measured by generally accepted financial data. These four platforms were chosen because of their use by contemporary businesses and their *suggested* impacts on business outcomes. It should be noted, however, that companies differ in their interpretations of social media and how that social media is accounted for (Nugroho, 2014).[1]

The companies that we included in this research were those on the 2013 Fortune 500 list. We chose these 500 companies because they are generally considered to be the most successful enterprises in the business world. Their social media activities and associated outcomes thus may contain more constructive implications that can be generalized than might be gleaned from small businesses. Also, as mentioned in Chapter 1, these companies represent a variety of industries/business sectors, and they tend to lead the trends in those industries/business sectors as well. In our research, we used the Global Industry Classification Standard to code each company into one of the ten following business sectors:[2]

1. Energy (e.g., Halliburton, Exxon Mobil)
2. Materials (e.g., Alcoa, Dow Chemical)
3. Industrials
   - Capital goods (e.g., Caterpillar)
   - Commercial and professional services (e.g., Kelly Services)
   - Transportation (e.g., Avis Budget Group)
4. Consumer discretionary
   - Automobiles and components (e.g., Ford)
   - Consumer durables and apparel (e.g., Nike)
   - Consumer services (e.g., Caesars Entertainment)
   - Media (e.g., DIRECTV)
   - Retailing (e.g., Target)
5. Consumer staples
   - Food and staples retailing (e.g., Supervalu)
   - Food, beverage, and tobacco (e.g., Dole Food)
   - Household and personal products (e.g., Avon)
6. Health care
   - Health care equipment and services (e.g., Abbott Laboratories)
   - Pharmaceuticals, biotechnology, and life sciences (e.g., Eli Lilly)
7. Financials
   - Banks (e.g., Bank of America)
   - Diversified financials (e.g., American Express)
   - Insurance (e.g., Aflac)
   - Real estate (e.g., CBRE Group)
8. Information technology
   - Software and services (e.g., Oracle)
   - Technology hardware and equipment (e.g., Apple)
   - Semiconductor and semiconductor equipment (e.g., Intel)
9. Telecommunication services (e.g., AT&T, Verizon)
10. Utilities (e.g., Duke Energy, Edison International)

As we have noted, our objective in conducting the research is to test the relationships between social media platforms and nonfinancial and financial outcomes for the Fortune 500 companies. At first glance, this sort of testing does not seem to be that complicated. For the financial side, it is not; financial data are easily available for companies that are publicly traded. For private companies, however, the data are more difficult to obtain. But for the social media there is much less consensus on *what* to measure and even less consensus on *how* to measure nonfinancial variables. In the following sections we will lay out the methods employed for data acquisition, beginning with companies' financial and accounting data as taken from two databases. We then turn to the acquisition of the public relations nonfinancial variables, the social media platforms used by companies, and how their key product indicators (KPIs) were operationalized. We will also address the sampling issue in our research and describe the statistical procedures implemented to test the relationships between financial and nonfinancial variables.

Financial Data Acquisition

The financial data were obtained from the Fortune 500 companies between January 2009 and December 2013 and covered a period of 5 years as reported through two databases: CRSP and Compustat. We used our university library subscription to access those two databases.[3] Specifically, we examined the financial outcomes for each social media platform on seven variables at two levels: financial (monthly stock return) and accounting (quarterly total revenue, net income or loss, earnings per share, profit margin, return on assets, and return on equity). First, the financial variable of interest was each company's monthly stock return, which we identified as the company's "success" on the stock market. We recorded each company's stock price information from 2009 to 2013 via the CRSP database based on its stock ticker. We also recorded the monthly return of the Standard & Poor's 500 Composite Index over the same time frame to control for stock market fluctuation. Second, there are accounting variables that may be influenced by nonfinancial variables that also impinge on "success" from a business performance perspective. Accordingly, we recorded each company's quarterly total revenue, net income (or loss), earnings per share, total assets, and total equity from 2009 to 2013 from the Compustat database. Each company's quarterly profit margin, return on assets, and return on equity were calculated based on the data. The specific formulas that we used in those calculations will be further explained in Chapter 4.

It should be noted that of the 2013 Fortune 500 list, 472 companies were publicly traded during our 5-year reporting frame. While most of those companies stayed public, some went private. Where we could obtain those companies' (incomplete) financial data from CRSP and Compustat, we included it. We must note, however, that 28 of the Fortune 500 companies are privately held and do not report all of the financial data used as the outcome variable in this study.

After collecting these financial data, we examined how the four social media platforms impact on companies with social media presences. In such explorations, the financial outcome variables of interest, including monthly stock return, quarterly total revenue, net income (or loss), earnings per share, profit margin, return on assets, and return on equity, were treated as dependent variables, the outcome of the predictors, or the independent variables (those variables that altered the outcome).

Nonfinancial Data Acquisition

The predictor variables represent the outcomes and outtakes obtained from observing and coding the variables across four social media platforms chosen both for their availability and for having been inferred to impact on company financial bottom lines. It should be noted that the companies chosen for study most often see their social media strategy as something greater than simply a return on investment (Kiron, Palmer, Phillips, & Berkman, 2013; Nugroho, 2014). Companies just building their social media typically justify the expenses in terms of return on investment. However, companies such as Dell or IBM have moved beyond ROI and recognize the larger role of social media in developing their internal and external communication strategies.

There are numerous social media platforms that have been used by the enterprises that encompass business. Over time, there is an ebb and flow of use and disuse. For instance, in the late 1990s the business world was taken by storm by social media platforms that allowed companies to create their own virtual worlds, engage in conversations with stock- and stakeholders, and actually create their own portals with currency and products (Edosomwan, Prakasan, Kouamc, Watson, & Seymour, 2011; Kaplan & Haenlein, 2010). This did not last long, however, as other more immediate forms of engagement came online. Besides the blog (really, nothing more than an extended text message), several social media platforms surfaced and retained their appeals and thus their impacts on business. We shall turn to the four employed in this book: Facebook, Twitter, YouTube, and Google+.

First, Facebook is the most established social media platform (set up in 2004). It has over 1 billion users, and although teens are reported to be using it less, it has a broad global reach. It is the go-to social media network for many, and it helps achieve awareness, knowledge, interest, and engagement. It is extremely interactive, with a large number of users commenting on other users' comments. Facebook's KPIs include likes, shares, comments, and comment valence.

Second, Twitter is a well-established social network that began more as a business-to-business network and morphed into more individual use. It was established in 2006 and has more than 250 million users and a fairly wide user audience. Twitter is limited to 140-character tweets, but the ability to point to imbedded source videos adds to its use. The platform is often used for real-time reporting of events or activities, and it helps build awareness, knowledge, and engagement. Twitter's KPIs include retweets, @replies, reply valence, favorites, and followers.

Third, YouTube is a platform for video sharing set up in 2005. It has over 1 billion users who can view, upload, and download videos of varying lengths. It helps to build awareness, knowledge (especially in terms of video instruction), and engagement. YouTube's KPIs include video views, likes (and dislikes), shares, comments, comment valence, and subscribers.

Finally, Google+ has a large international audience, especially among technology early adaptors. It was developed by Google in 2011 to directly compete against Facebook. It is both text- and video-based and issued most to build awareness and add knowledge. Google+'s

KPIs include plusses, shares, comments, comment valence, and followers.

In order to record the KPIs from each social media platform, we need to take some sampling issues into consideration. From a statistical standpoint, there is a crucial difference between *sample* and *population*. Here, the term *population* does not necessarily mean human beings. It actually refers to the whole set of units of research interest. A *sample* is a subset of the population. The population is up to the researcher to define, and the sample should be selected based on a certain criterion and method. For example, in our Facebook research we defined all Facebook posts created by the Fortune 500 companies between 2009 and 2013 as the population. As it comes to a sample, we then need to figure out how many posts we should collect for our research (sample size) and how to collect them (sampling method). The whole point of statistical analyses (i.e., inferential statistics) is to examine the sample data and then generalize the findings to the population.

Because the financial data in our study were measured on a monthly or quarterly basis, we decided to collect one Facebook post from each company for each month as long as the company had at least one post in that particular month. In this way, we could then match each company's financial data and nonfinancial data together, month by month and quarter by quarter. Regarding the specific sampling method, we took two steps in selecting a post if a company had multiple posts in a given month (if the company had only one post in that month, of course that post was selected). First, we used an online random number generator to generate a number between 1 and 30 or 1 and 31, depending how many days a particular month had. For example, if the random number was 5, we then looked at the fifth day of that month and found out whether the company had posted anything on that day. If the company had just one post on the day, then that post was selected. If the company had more than one post on the day, we then used the random number generator again to decide which post to pick. If the company had no post on the day, we then looked at the closest day (e.g., the fourth or the sixth in this case) when the company had a post. After we selected each sample post, we then recorded the KPIs associated with it, such as the numbers of likes, comments, and shares. To gauge the comment valence, we also recorded the top two comments on each sample post. We repeated these procedures for all four social media platforms. As the readers can tell, based on this description, obtaining nonfinancial data from social media is much harder and also more troublesome than obtaining financial data.

Nonfinancial Variables

As noted throughout this volume there are variables other than financial that have demonstrated business outcomes. We noted in Chapter 2 that nonfinancial variables can demonstrate impact on business outcomes but are more difficult to define and hence measure. Nonfinancial variables deal with how stock- and stakeholders view a company. That view, the result of a communications plan, yields "return on expectations," or ROE (Stacks, 2002). The key is stakeholder *relationship* (for the moment, let's consider the stockholder as a special type of stakeholder, one who invests in the company and thus has a stake in how it is managed and how it is perceived by other stakeholders who might affect it, such as customers, employees, analysts, and so forth). Roger Hayes (2014), a business consultant with a considerable international reputation, suggests that businesses should be looking at what he calls "stakeholder primacy," with the focus on expanding stakeholder relationships, thus increasing company trust, credibility, and authority through transparent leadership and communication with multiple stakeholders. The downside of this transparency, however, is that the more transparent a company is, the more it may actually induce crises unless communication is carefully managed.

Given the increased focus on social media relationships, immediate feedback, and a quickly moving network, the success and failure of a company is now often defined more by its social media presence and less by its presence in the traditional media (e.g., newspapers, television, radio). What social media allow a company to do is to create immediate and often deeply connected networks among its stakeholders. The outcome of this is to create, establish, and nurture the company's nonfinancial variables that in turn impact on financial and business outcomes.

In several publications Stacks and his associates (Michaelson & Stacks, 2014; Stacks, 2011; Stacks & Michaelson, 2010) have argued that public relations has direct responsibility in business communication in terms of creating the company's *credibility*—believability in the eyes of stakeholders that the company will do as it says it will do. A company with high credibility is one that is listened to, seen as representing something the stakeholder believes is meant for him or her, or seen as an authentic entity that backs its values with actions. In fact, public relations has been defined as the "management of [a company's] credibility" (Stacks, 2002, 2011; Michaelson & Stacks, 2014).

Closely related to credibility is *reputation*. Reputation is historical in that what a company has done in the past is seen as representing what it might do in the future. As such, reputation plays a critical role in business, and is often one of the first casualties when a company has a crisis of some sort, whether it is due to management failure, technological failure, malfeasance, or some other cause. How important is reputation? Warren Buffet is reported to have said, "It takes 20 years to build a reputation and 5 minutes to ruin it. If you think about that, you'll do things differently" (Goodman, 2013).

Another important nonfinancial variable is *trust*. Trust is as future-oriented as reputation is historical. Trust comes from a relationship whereby open two-way communication occurs, what public relations guru James Grunig and his associates (Grunig & Grunig, 1992; Grunig, Grunig, & Dozier, 2002) have called "two-way symmetrical communication." In the traditional sense of media relations, communication was historically one-way; it took a mass-mediated message and let it loose on a large, targeted "public." There was no relationship involved, and credibility was assumed to come from the message source (e.g., newspaper, television, advertisement). This then transformed into two-way "asymmetrical" communication, whereby after the message was transmitted, the public's perceptions of the object of the communication were assessed. This usually occurred weeks after transmission of the message. Thus, little or no trust was established. Two-way symmetrical messages engage the stakeholder to comment on the company, even if that commentary is negative, in the belief that an open and honest dialogue will result in stronger ties to the company—stronger ties that yield increased profits from a relationship built on trust.

Finally, we have *confidence*. Confidence is a feeling by stakeholders that a company will actually do what it says it will do, its product actually works as promised, and so forth. If a stakeholder has confidence in the market, for instance, he or she will invest more money and more often. However, even with a company that is credible and trusted, and that has a reputation for business success, if confidence is low, then investment in its stock or purchases of its product, or in the case of the employee, quality workmanship, may suffer.

Clearly, these five nonfinancial variables are intercorrelated. As with much human behavior, decisions are not made in separate decision-making stages, but are often intertwined as expectations from one variable to the next change with circumstances, times, and conditions. However, we can operationally define the key product indicators from our four social media platforms as discrete units of analysis and observe how over time they have impacted on financial and business success.

Data Coding and Data Analyses

As noted earlier, financial data was obtained from publicly available documentation. The data were then entered into a statistical program, IBM-SPSS v. 22, for data analysis, interpretation, and evaluation. The next five chapters will present this in more detail, but basically, each chapter looks first at how the companies use the social media platform and then, second, how that platform and its KPIs relàte to financial and accounting outcomes.

The social media variables were operationalized for you earlier. Each social media platform's communications were observed for several months in the earlier part of 2014. The social media messaging was observed by independent coders who, after training, were able to code the messages at 90% reliability or higher, thus chance error was minimal.[4] Each of the following four chapters goes into greater detail about the coding and the KPIs selected due to differences in access to data and the types of data that had to be coded. In total, we had eight coders. They were trained to code just one or two of the four social media platforms; this increased reliability and prevented coder fatigue that would have occurred if only two coders had attempted to code all the data collected.

The KPIs chosen represent what the social media research industry has used previously. Although we acknowledge that some KPIs may be better than others, the scope and nature of this study prompted us to use traditional social media metrics associated with each platform. For Twitter, the *tweet* was the unit of analysis. The data were coded specifying what kind of tweet it was (reply, retweet, favorite), and the tweet's valence was evaluated as positive, neutral, or negative. A company's presence on Twitter was also coded for total number of tweets, total number of people who were following the company, and total number of people the company is following. For Facebook, each *post* was coded specifying whether it was liked, had a comment, and/or was shared, and the comments' valences were evaluated (positive, neutral, negative). For YouTube, each *video* was coded specifying views, likes, dislikes, comments, video lengths, shares, and comments' valences (positive, neutral, negative). Each company's general YouTube activity was coded for total number of subscribers, total number of views, and total number of videos. Finally, for Google+ each *post* was coded in terms of comments, plusses, shares, and the comments' valences (positive, neutral, negative). Regarding the general activity on Google+, each company was coded for total number of followers, total number of views, and total number of people in the company's circle.

In the SPSS database, each of the company's KPIs was treated by social media type as nonfinancial data. In addition, five nonfinancial variables were computed. In particular, a gross metric of *relationship* was operationalized as total number of Twitter replies, total number of Facebook comments, total number of YouTube comments, and total number of Google+ comments. A composite relationship variable was the sum of the social platform KPIs divided by four. *Reputation* was determined by the total numbers of Twitter favorites, Facebook likes, YouTube likes, and Google+ plusses. The composite reputation variables was the sum of each platform's KPIs divided by four. *Trust* was operationalized as the total number of Twitter retweets, total number of Facebook shares, total number of YouTube shares, and total number of Google+ shares. Again, the composite variable was the sum of each platform's KPIs divided by four. *Credibility* was computed as total number of Twitter followers, total number of YouTube subscribers, and total number of Google+ followers.[5] The composite variable was the sum of each platform's KPIs divided by three. Finally, *confidence* was operationalized as the Twitter reply overall valence, the Facebook comment overall valence, the YouTube comment overall valence, and the Google+ comment overall valence. The composite variable was computed by summing the platform's valence and dividing by four.

Clearly, some may object to the way we have defined the KPIs coded to produce our nonfinancial variables. It should be understood, however, that this study is more of a field experiment method with the data defined based on its use. In this way we have approached the process of defining variables by KPIs and then associating them with financial and accounting data. Take this as a caveat before you turn to the next chapter and evaluate the findings within the confines of the data and method.

The following five chapters will assess the findings statistically. We arranged the order of Chapters 4 to 7 as Twitter, Facebook, YouTube, and Google+, based on their adoption rates by the Fortune 500 companies.[6] Each chapter begins with an overview of the social media platform and a summary of how the data were collected. Each then presents the platform's KPIs findings as a descriptive indicator of use. The chapters then turn to examining the relationship KPIs and nonfinancial variables have to financial and accounting outcomes. This analysis is conducted through multiple regressions and is presented in two ways:[7] with a simple description of the statistical method and the results, and, in endnotes, with more detail for readers with in-depth knowledge of the statistical tools employed. After the social media platform chapters have been presented, a composite variable chapter (Chapter 8) will look at the effects each of the nonfinancial variables has on the financial and accounting data. Taken together, the analyses presented in all five chapters aim to answer the following research questions:

- How do the Fortune 500 companies use social media to communicate with their publics in general?

- How does the public engage in social media to communicate with the Fortune 500 companies?
- How do we measure a company's social media activeness in terms of confidence, credibility, relationship, reputation, and trust?
- Does a company's social media nonfinancial activeness affect its business performance from a finance perspective, measured in its monthly stock return?
- Does a company's social media nonfinancial activeness affect its business performance from an accounting perspective, measured in its quarterly total revenue, net income (or loss), earnings per share, profit margin, return on assets, and return on equity?
- Do companies in different business sectors use social media differently?
- Are there significant trends in using social media over the years from 2009 to 2013?

Summary

After reading this chapter the readers should have an understanding of the variables employed and how they were operationalized and created in the study. The next five chapters will explore in depth the relationships between nonfinancial variables and financial and accounting outcomes. We believe that some of the findings will surprise you, others will reinforce what you may have already surmised, and some take us to places we had not intended to go. Regardless, each chapter expands our knowledge of how social media variables impact company financial and accounting variables. We might summarize by noting that it is "a brave new world" out there in social media space, and this is the first attempt to quantify the new world's impact on business success.

Chapter Four

Twitter

Chapter Overview

This chapter describes how the Fortune 500 companies used Twitter to communicate with their publics in a 5-year period from January 2009 to December 2013. Based on the statistical analyses of data collected from Twitter and other resources, we aim to address the following questions:

- Is Twitter a viable vehicle for companies to use for corporate communication purposes?
- How does the public respond in general to companies' communication messages on Twitter?
- How can a company's nonfinancial activeness on Twitter be measured?
- Is a company's Twitter nonfinancial activeness associated with its business performance from a finance perspective, measured in the monthly stock return?
- Is a company's Twitter nonfinancial activeness related to its business performance from an accounting perspective, measured in

the quarterly total revenue, net income (or loss), earnings per share, profit margin, return on assets, and return on equity?

- Do different business sectors differ in their nonfinancial activeness on Twitter?

Companies' Activities on Twitter

The raw data of how the Fortune 500 companies used Twitter were collected from January 22 to April 2, 2014. During this data collection period, a total of 420 companies on the Fortune 500 list were found to have an account on Twitter. Among those 420 corporate Twitter accounts, 181 of them were verified by the companies. There were three major indicators to suggest how actively a company communicated nonfinancially with their publics on Twitter: (1) the number of "tweets," (2) the number of "following," and (3) the number of "followers." Thus, we recorded each company's total numbers of tweets, following, and followers on a given day during the data collection period. It was found that the companies' activities on Twitter varied to a large extent. Some companies had no tweets at all, whereas some other companies tweeted heavily. On average, the companies posted 8,055 tweets on their accounts (minimum: 0; maximum: 445,699). Those companies that appeared to be relatively inactive on Twitter seemed to have fewer following and followers, while others that tweeted more frequently tended to have more following and followers. On average, the companies had 4,317 following (minimum: 0; maximum: 543,802) and 192,835 followers (minimum: 1; maximum: 13,621,880). Facebook, Google, and Starbucks were the three most popular companies on Twitter, having 13,621,880; 7,923,826; and 5,795,263 followers, respectively.

To test whether companies' Twitter nonfinancial activities were significantly associated with their Fortune 500 rankings (for this book, we used the 2013 Fortune 500 rankings), we conducted a correlation analysis. To ensure that statistical jargon does not hinder this book's readability, we decided to explain in layman's terms in this and later chapters the statistical analyses we performed, and then, where applicable, to include in the chapter endnotes specific details of those statistical tests for the readers' reference.[1] In general, the purpose of a correlation analysis is to test whether one thing's change can be reflected in another thing's change, either positively or negatively. We found a significant correlation between a company's Fortune 500 ranking and its total number of tweets: *Companies that ranked higher on the Fortune 500 list posted more tweets on Twitter.*[2] This meant that the number of tweet messages that a company delivered to its publics on Twitter reflected the company's size and overall reputation. However, we did not find any significant relationship between a company's Fortune 500 ranking and its total numbers of following or followers.

We also conducted a correlation analysis with the three Twitter indicators: the number of tweets, the number of following, and the number of followers. They were all significantly and positively correlated together. That is to say, *a company that tweeted more on Twitter tended to have more following and followers as well.*[3]

Companies' Tweets

As described in Chapter 3, we collected up to 60 random tweets from each company that had a corporate account on Twitter from January 2009 to December 2013, one tweet for each month. We believe such a sample could reasonably represent how each company communicated with their publics in the 5-year period, although it is not perfect. On the one hand, for those companies that tweeted thousands of times, their communication messages might be underrepresented by this sample. On the other hand, for those companies that rarely tweeted, their Twitter activities might be overrepresented. However, since we performed this systematic *sampling* procedure with each company and examined the data at an aggregate level, the individual differences across different companies would likely be cancelled out in the data analyses. Also, we purposely sampled in this way so that we could test the relationship between a company's Twitter nonfinancial activeness and its business financial performance, because a publicly traded company's business performance is typically evaluated on a quarterly basis. Finally, we found this sampling method to be more feasible and appropriate compared to other options. For example, if we had adopted the simple random sampling method (in general, simple random sampling means to randomly select a certain number of units out of all units available) and selected a random 10% of each company's tweets, we would have collected "nothing" from some companies (those that had fewer than 10 tweets) but "a lot" from others (for example, we would have needed to collect 19,585 tweets from Wal-Mart). By doing so, the data undoubtedly would have been heavily skewed toward the companies that had a more active presence on Twitter.

Once we sampled a tweet from a company, three metrics associated with this tweet were recorded: (1) the number of "replies," (2) the number of "retweets," and (3) the number of "favorites." We also recorded the top two replies to this tweet if it generated at least two replies (if the tweet generated only one reply, we then recorded that lone reply; if the tweet generated no reply, we then considered its reply data as none [0]). Again, this sampling method was imperfect, but the sample could reasonably represent how the public responded to a company's communication messages on Twitter. This systematic sampling was also more feasible compared to other sampling options (for example, when a tweet generated "too many" replies, it would be extremely difficult to do the simple random sampling).

Using this method, we collected a total of 9,122 tweets, representing 420 companies' Twitter activities from January 2009 to December 2013. On average, each tweet generated 0.43 replies (minimum: 0; maximum: 39) and 3.86 favorites (minimum: 0; maximum: 1,279), and was retweeted 7.96 times (minimum: 0; maximum: 3,935). We conducted a correlation analysis with these three metrics and found that they were significantly correlated with each other.[4] *When a tweet was retweeted more often, it also tended to generate more favorites and replies.*

We also collected a total of 1,509 replies to those tweets, using the method described earlier. To further understand how the public responded to companies' tweets in those replies, we implemented a content analysis procedure. Content analysis is a quantitative research method widely adopted in the social sciences. Its main purpose is to provide an objective description of a certain set of "content." Specifically in this case, our content referred to the 1,509 replies. To analyze this content, we engaged two graduate students at our university to code each reply into one of the five categories: (1) **compliment**—the replier provided a mostly positive opinion toward the tweet, (2) **complaint**—the replier provided a mostly negative opinion toward the tweet, (3) **question**—the replier posed a question in relation to the tweet, which was neither positive nor negative, (4) **self-promotion**—the replier provided a piece of information that had little to do with the tweet, with the purpose of promoting himself or herself, and (5) **neutral opinion**—the replier provided a somewhat neutral opinion toward the tweet. The reason why we had two coders involved in this process was to ensure that their coding was independent and unbiased. Regarding the specific procedure, we adopted a traditional two-step content analysis approach: (1) reliability tests, and (2) coding. First, we explained the conceptual definition of each category to the two coders in a training session and provided them with real examples. We also encouraged them to ask questions to ensure that there was no ambiguity. Then we randomly selected five companies that had corporate Twitter accounts and asked both coders to independently code the replies to those companies' tweets. After they completed this task, their coding was compared. The agreement percentage of coding between the two coders was 92.8%, suggesting a satisfactory level of reliability.[5] For those replies where the two coders disagreed with each other, we resolved the discrepancy by discussion. Finally, the rest of the replies in our sample were evenly split, and each coder was asked to code half of them. The whole coding process lasted approximately 4 weeks. For illustrative purposes, a real example for each coding category is listed below (all examples are from our sample, and we list them here without correcting any spelling or grammar) so that the readers can further understand how the categories differ from one another.

(1) Compliment: "Love it! I will say it!"
(2) Complaint: "What is this? Learning propaganda at a young age? I hope you are teaching disaster cleanup."
(3) Question: "will hotchips slides be posted somewhere?"

(4) Self-promotion: "1202 Christmas Holiday SUPERDEALS The Deepest DISCOUNTS on Top Quality" (to promote an online shopping website that was not the company that posted the original tweet)
(5) Neutral opinion: "i will be a potatoe" (to answer the tweet question "Have you decided what you are going to be for Halloween?")

Overall, the replies to the companies' tweets appeared to be mostly positive or neutral (31.2% compliment, 14.5% complaint, 9.7% question, 3.6% self-promotion, and 41.0% neutral opinion). In order to test how these replies might be associated with the companies' business performances, we conducted the following data transformation in our database to reflect the valence of each reply: We recoded a compliment as +1 because it had a positive connotation, a complaint as –1 because it carried a negative meaning, and everything else as 0 because they were all somewhat neutral. Since there were up to two replies associated with each tweet in our database, we averaged the scores of the two replies whenever applicable. For example, if a tweet generated two replies with one being a compliment and the other being a neutral opinion, the average reply valence score for this tweet would be (1 + 0)/2 = 0.5. Based on such a calculation method, each tweet's average reply valence score would be in the range of –1 to +1, with –1 representing the worst case of third-party endorsement and +1 representing the best scenario of third-party endorsement. Given that each tweet generated a different number of replies, we further calculated each tweet's overall reply valence using the following formula:

Overall reply valence = number of replies × average reply valence

For instance, if a tweet generated 7 replies, and its average reply valence was 0.5, then its overall reply valence would be 7 × 0.5 = 3.5. We need to point out, however, that this score is a rough estimation of each tweet's overall reply valence; it is not 100% accurate because we sampled only the top two replies to each tweet.

Companies' Twitter Activities and Their Stock Returns

One of the major objectives of this book is to test the relationship between a company's social media activities and its business performance. Having systematically collected sample tweets and replies from those companies that had accounts on Twitter, we wanted to match their Twitter nonfinancial activeness to their performance in the real business world, month by month, and quarter by quarter. To fulfill this goal, we included two business perspectives in our examination: *finance* and *accounting*. On the one hand, from a finance perspective, a company's "success" could be evaluated based on its stock return. On the other hand, the judgment of a company's business performance from an accounting perspective would be centered around its return on assets and return on equity. As described in Chapter 3, there were a total of 472 companies on the Fortune 500 list that were public for a period between January 2009 and December 2013, so their business performance data from both the finance and accounting perspective were accessible. Most of those companies stayed public during the whole 5-year period, but a few did not. For example, Dell went from public to private in late October 2013.[6] In such cases, the company's financial data were still usable though they were not complete for the whole 5 years. Therefore, we included all companies' business performance data in our statistical analyses as long as they were obtainable from the public databases.

Specifically, we used each of the 472 public companies' stock tickers to search for its stock price information from January 2009 to December 2013 in the database *CRSP*. When calculating a company's monthly stock return, its closing price on the last trading day of the current month was compared to that of the previous month. For example, the closing stock price of Johnson & Johnson (stock ticker: JNJ) on the last trading day of October 2013 (October 31) was $92.61, and its closing price on the last trading day of September 2013 (September 30) was $86.69. Thus, the monthly return of Johnson & Johnson on the stock market in October 2013 could be calculated as follows:

$$JNJ's\ October\ 2013\ stock\ return = (92.61/86.69) - 1 = 1.0683 - 1 = 0.0683$$

However, we need to point out that the calculation above does not adjust for any stock split (or reverse stock split) or dividend. To ensure that the calculation of all companies' monthly stock returns were adjusted appropriately and consistently, we recorded the "Holding Period Return" in the database CRSP for each of the 472 companies (the "holding period" in this case was equal to 1 month, and it was adjusted for stock splits/reverse stock splits and dividends) and used those data for our later statistical analyses.[7]

Because each company's stock price was likely to be affected by the fluctuation of the whole stock market, we also recorded the monthly return of the Standard & Poor's 500 Composite Index from January 2009 to December 2013, so as to be able to control such a factor in our statistical analyses.[8] This stock market index monthly return information was obtained from the database CRSP (the "Return on S&P Composite Index"), based on a comparison of the Standard & Poor's 500 Composite Index at the end of the current month to that of the previous month.[9] For example, the Standard & Poor's 500 Composite Index for September and October 2013 was 1681.55 and 1756.54, respectively. The monthly return on S&P Composite Index in October 2013 could thus be calculated as follows:

$$S\&P\ 500's\ October\ 2013\ return = (1756.54/1681.55) - 1 = 1.0446 - 1 = 0.0446$$

To figure out whether the Fortune 500 companies' Twitter activeness was significantly associated with their stock returns, we performed a statistical procedure called "hierarchical regression." In general, the purpose of such an analysis is to test whether one thing's change can cause another thing to change, after controlling the effect of a third thing. In our case, we wanted to examine whether the change in a company's Twitter nonfinancial activeness could "predict" the change of its stock return, independent of the effect of the whole stock market on the company's stock price.[10] Specifically, we had six variables in our database, as follows:

(1) The monthly stock return of each public company,
(2) The monthly return of the Standard & Poor's 500 Composite Index,
(3) The number of replies to each tweet (*relationship*) (it represented a specific month when the tweet was created on Twitter),
(4) The number of retweets (*trust*) of each tweet (it represented a specific month when the tweet was created on Twitter),
(5) The number of favorites (*reputation*) of each tweet (it represented a specific month when the tweet was created on Twitter), and
(6) The overall reply valence score (*credibility, confidence*) of each tweet (it represented a specific month when the tweet was created on Twitter).

Given that the monthly stock return of each company was significantly correlated with that of the Standard & Poor's 500 Composite Index,[11] and that there were significant correlations among each tweet's numbers of replies, retweets, and favorites, and its average reply valence,[12] we standardized all six variables in our database in order to avoid any potential "collinearity" problem in the regression analyses.[13] We then tested the relationship between the companies' Twitter activeness as the nonfinancial variables of interest and their stock returns as the financial variable of interest by hierarchical regression. The regression equation can be represented as follows:

Equation 1: $y = a + b_1x_1 + e$, *and*
Equation 2: $y = a + b_1x_1 + b_2x_2 + b_3x_3 + b_4x_4 + b_5x_5 + e$, *where*

y = the monthly stock return of each public company
a = constant
x_1 = the monthly return of the Standard & Poor's 500 Composite Index
x_2 = the number of replies to each tweet
x_3 = the number of retweets of each tweet
x_4 = the number of favorites of each tweet
x_5 = the overall reply valence score of each tweet
b_1 to b_5 = regression coefficients
e = residue

As seen in these two equations, we aimed to test the effects of x_1, x_2, x_3, x_4, and x_5 on y. Specifically, we entered x_1 and y into the first equation because the effect of x_1 on y needed to be controlled. As expected, b_1 appeared to be statistically significant, suggesting that the companies' monthly stock returns were largely affected by the fluctuation of the Standard & Poor's 500 Composite Index.[14] Next, we entered x_2, x_3, x_4, and x_5 into the second equation. By comparing the two equations, we would be able to find out if any of those four Twitter variables (x_2 to x_5) had an "additional" or "incremental" effect on y, independent of the influence of x_1 on y. The analyses results showed that none of the four coefficients (b_2 to b_5) was statistically significant. In other words, *we did not find any significant effect of each tweet's number of replies, retweets, or favorites, or overall reply valence on each company's monthly stock return.*[15]

Companies' Twitter Activities and Their Profitability

In addition to testing the relationship between a company's Twitter nonfinancial activeness and its stock return, we also intended to examine whether a company's Twitter activeness was associated with its *profitability*, primarily from an accounting perspective. We used several mainstream accounting measures to gauge a company's profitability in this book, including the total revenue, net income (or loss), earnings per share, profit margin, return on assets, and return on equity. Specifically, we used each company's stock ticker to search for the report on its quarterly total revenue, net income (or loss), earnings per share (diluted and including extraordinary items), total assets, and total equity from January 2009 to December 2013 in the database Compustat. We also recorded what business sector each company belonged to in the same database, according to the Global Industry Classification Standard. After we obtained such information, we calculated each company's quarterly profit margin, return on assets, and return on equity. The calculation formulas we used are listed as follows:

(1) *Profit margin = Net income (or loss)/Total revenue*
(2) *Return on assets = Net income (or loss)/Total assets*
(3) *Return on equity = Net income (or loss)/Total equity*

To match a company's Twitter activeness with its profitability quarter by quarter, we calculated each company's Twitter activeness on a quarterly basis. For example, the replies a company generated in January, February, and March were added up to represent its total replies in the first quarter of each year. Similarly, the number of retweets, the number of favorites, and the overall reply valence score were added up accordingly. However, when we recorded the data from the database Compustat, we noticed that most companies (76.2%) chose to end their fiscal year in December. For those companies that did not end their fiscal year in December, we calculated their quarterly Twitter activeness based on whatever month they ended their fiscal year. For instance, if a company ended its fiscal year in January, the replies it generated in February, March, and April were added together to represent its total number of replies for the first quarter of that year.

Before we conducted the regression analyses, we created a few "dummy variables" for the purpose of statistical control.[16] Specifically, we created three dummy variables for seasonality (because there were four seasons each year) and nine dummy variables for business sector (because there were 10 business sectors based on the Global Industry Classification Standard). The purpose of these dummy variables in the regression equation was to control for the impact of seasonality and business sector on a company's profitability. It was not our focus to determine whether seasonality might help companies generate more profits or what particular business sector(s) tended to have a higher level of profitability. Thus, we wanted to control for such influences in our statistical analyses and be able to explore the independent effect of a company's Twitter activeness on its profitability. In summary, we did a series of hierarchical regression analyses, testing the effect of a company's Twitter activeness on its profitability from six different business perspectives. The regression equation could be understood as follows (all the variables were standardized in the analyses except for the dummy variables):

Equation 1: $y = a + b_1x_1 + \ldots + b_3x_3 + b_4x_4 + \ldots + b_{12}x_{12} + e$, *and*
Equation 2: $y = a + b_1x_1 + \ldots + b_3x_3 + b_4x_4 + \ldots + b_{12}x_{12} + b_{13}x_{13} + b_{14}x_{14} + b_{15}x_{15} + b_{16}x_{16} + e$, where

y = each public company's quarterly total revenue, net income (or loss), earnings per share, profit margin, return on assets, or return on equity
a = constant
x_1 to x_3 = three dummy variables for seasonality
x_4 to x_{12} = nine dummy variables for business sector
x_{13} = the quarterly number of replies
x_{14} = the quarterly number of retweets
x_{15} = the quarterly number of favorites
x_{16} = the quarterly overall reply valence score
b_1 to b_{16} = regression coefficients
e = residue

The results of the regression analyses suggested several significant effects. In particular, the number of retweets appeared to be a strong indicator of the companies' business performances. Specifically, *a company's quarterly number of retweets showed a significant and positive effect on its quarterly net income (or loss), profit margin, and return on assest.*[17] In addition, *the number of favorites had a significant but negative impact on the return on assets.*[18] Finally, we did not find any significant effect of the number of replies and the overall reply valence on any of the business performance measures.[19]

The first significant finding of these regression analyses is easy to understand and interpret. The number of retweets of a company's nonfinancial messages on Twitter was *positively* associated with the company's reported net income (or loss), profit margin, and return on assets, probably because the number of retweets suggested the intensity of attention that the company drew from the public during a particular period. However, the second significant finding, that the number of favorites was *negatively* related to a company's return on assets, looked quite counterintuitive. There are several possible explanations for such a finding. First, it is worth noting that each tweet's number of favorites was negatively associated with its average reply valence (see endnote 12). According to the *negativity bias* theory in the persuasive communication literature (e.g., Chen & Lurie, 2013; Eisend,

2010), people generally consider negative information as more credible than positive information. In other words, negative messages may look more genuine and trustworthy. In this particular case, we suspect that the public might consider a company's tweet to be their "favorite" when the replies to it appeared to be somewhat negative because they saw a higher level of "openness" or "honesty." Thus, a larger number of "favorites" might suggest, counterintuitively, that there were more people criticizing the company than praising it. Second, the finding could be caused by a few "outliers" in the data. After excluding those outliers in a separate regression analysis, the significant effect of the number of favorites on the return on assets disappeared.[20] Third, this finding might be caused by limitations of our data. It is possible that the regression analyses presented a misleading result because the data were somewhat biased. We will further explain what shortcomings our data had in a later section, and we suggest that readers should interpret this finding with caution.

Finally, we did a series of independent t-test analyses to compare the companies' Twitter activeness when they were profitable to when they were not.[21] Specifically, we created a new variable in our database called *profitability*, based on each company's quarterly reported net income (or loss). When a company reported a net loss (a negative figure) for a specific quarter, we considered it to be a nonprofitable quarter for this company and coded it as 0. In contrast, when a company reported a net income (a positive figure) for a particular quarter, we considered it to be a profitable quarter for this company and coded it as 1.[22] We then conducted t-test analyses to compare these two types of quarters (a company could be profitable in one quarter but not in another, thus this comparison is between two types of quarters instead of two types of companies) on the numbers of replies, retweets, and favorites, and the average reply valence. It was found that *a company's tweet was retweeted significantly more times when it was profitable* (approximately 28 times on average) than when it was not (approximately 12 times on average).[23] No other significant difference was detected between profitable and nonprofitable quarters.[24]

Companies' Twitter Activities Across Business Sectors

As described earlier, we classified the Fortune 500 companies into 10 different business sectors based on the Global Industry Classification Standard, including energy, materials, industrials, consumer discretionary, consumer staples, health care, financials, information technology, telecommunication services, and utilities. To examine whether different business sectors differed in their Twitter activeness, we performed a series of one-way ANOVA tests.[25] In particular, we compared those 10 business sectors in terms of the number of replies, the number of retweets, the number of favorites, and the average reply valence, on a quarterly basis.

We found several significant results. First, there existed significant differences among the 10 business sectors in regard to the number of replies.[26] *Telecommunication services was the "winner" in generating replies,* based on a post hoc analysis, because its average number of replies (approximately 10.24) was significantly higher than any other sector's (ranged from 0.28 to 2.62).[27]

Second, the *consumer discretionary sector generated a significantly higher average number of retweets* (approximately 69.11) than all other sectors except for information technology, consumer staples, and telecommunication services (ranged from 2.23 to 9.77), based on our ANOVA test and post hoc analysis.[28]

Third, the *consumer discretionary sector appeared to be the most successful in generating favorites,* too. The average number of favorites generated by the consumer discretionary sector (approximately 38.67) appeared to be significantly higher than all other sectors except for information technology and telecommunication services (ranged from 0.54 to 12.87), according to the ANOVA test and post hoc analysis.[29]

Finally, we compared the 10 business sectors regarding their average reply valence. The ANOVA test and post hoc analysis did not suggest a single "winner" or "loser." There was *no single sector that generated significantly more positive or negative replies than all other sectors*. However, we did find that the *average reply valence of consumer staples and consumer discretionary was significantly more positive than that of utilities, energy, and telecommunication services.*[30]

Limitations of Our Data

Before we summarize our findings presented in this chapter, we need to point out that there were certain limitations associated with our data. The biggest shortcoming of the data was that it did not fully represent some companies' Twitter activities throughout the whole 5-year period. The reason for this was that although we recorded the data from each company's Twitter account, the Twitter website allowed each account to present only up to 3,200 tweets.[31] Therefore, if a company had fewer than 3,200 tweets in its entire tweeting history, our sample tweets well represented its activities, but for those companies that were highly active on Twitter, such as Wal-Mart, Whole Foods Market, and Foot Locker (their total number of tweets was way over 3,200), our data represented their Twitter activities for only a few months in 2013. Since social media data can change every day, or even every minute, it is very difficult to find a "real" random sample in social media research. Our systematic sampling method for this platform could reasonably represent the Fortune 500 companies' use of Twitter at an aggregate level, but we do want the readers to be aware of its shortcomings.

We also want to point out that our content analyses of the tweets' replies generally reflected the valence of the public's responses to the companies' nonfinancial messages on Twitter, but it did not show different degrees of intensity or *tone*. For example, when a consumer complained about a company in his or her reply to the company's tweet, the complaint could be "just a little bit" negative or "extremely" negative. Our coding did not touch upon such differences. This limitation applies to the coding in all four social media platforms included in our research. In the content analysis process, we decided to simplify rather than complicate the coding categories for our coders because we wanted to balance accuracy and efficiency. Given that the coders were requested to code thousands of pieces of information in a given time, we wanted to make the coding categories as clean as possible to avoid confusion and fatigue. However, this did bring certain limitations to our data analyses.

Finally, we need to clarify to the readers how they should interpret the insignificant results reported in this book. When we performed the statistical analyses, we used the traditional criterion in social science research to determine significance ($\alpha = .05$). When a p-value was larger than .05, it would lead to a conclusion of statistical insignificance. Generally speaking, significance suggests existence of a tested relationship, but insignificance cannot "prove" the nonexistence of that relationship. This argument sounds confusing but it is important to understand. We will use a metaphor here to help illustrate this point. Suppose that a person goes to a lake for fishing, and he successfully catches a fish after spending an hour there. In this case, it is "significant" and the conclusion is that the lake has fish (because the person catches a fish, based on logic, the lake must have at least one fish). However, in another case, if the person goes to the lake but fails to catch any fish after an hour, it is "insignificant" but the conclusion cannot be that the lake has no fish (because there are many potential reasons other than that the lake has no fish to explain why he catches nothing, such as that the weather is bad for fishing, or that his fishing skills are not good enough). To apply this logic to our findings reported in this chapter, although we did not find a significant relationship between a company's Twitter activeness and its monthly stock return, we cannot conclude that what a company does on Twitter has nothing to do with its stock return. It was possible that this insignificance was caused by the limitations of our data.

Synopsis of Research Findings

As noted earlier, we recorded the raw data from multiple sources for this Twitter research. The database that we set up contained a large amount of information and we analyzed the data from a wide variety of angles by using different statistical procedures. It is necessary for us to synopsize the key findings from this project:

- Twitter seemed to be a popular platform for the Fortune 500 companies to communicate with their publics. During our data collection period, we found that 420 companies on the Fortune 500 list (84%) had an account on Twitter.

- The usage intensity of Twitter by each company varied to a large extent. Some companies did not tweet at all, and some others tweeted fairly frequently.
- A company's total number of tweets as a measure of *relationship* was positively associated with its total numbers of following and followers on Twitter.
- The public showed a reasonable level of *relationship engagement* with the companies' communication messages on Twitter, as measured by the numbers of replies, retweets, and favorites.
- When the public replied to a company's tweet, it was usually positive or neutral.
- We did not find any significant relationship between a company's Twitter nonfinancial activeness and its monthly stock return.
- We found that the number of retweets (*trust*) was significantly and positively associated with the net income (or loss), profit margin, and return on assets, measured on a quarterly basis. That is to say, when a company had more retweets, it tended to have a higher level of profitability.
- We also found a significant and negative association between the number of favorites (*reputation*) and the return on assets. This finding was counterintuitive and there are multiple possible explanations. We suggest that readers interpret this finding with caution because this effect disappeared after we reconducted the data analysis excluding a few outliers in the database.
- We did not find any significant relationship between the number of replies (*relationship*) and the six profitability measures. The overall reply valence (*confidence*) did not show any significant impact on the profitability measures, either.
- A company's tweet tended to be retweeted significantly more times during a profitable quarter than during a nonprofitable quarter (*trust*).
- Different business sectors differed in their Twitter activeness. First, telecommunication service companies tended to generate significantly more replies than other businesses. Second, consumer discretionary companies appeared to be the most successful in generating retweets and favorites. Finally, the public's

replies to the tweets of consumer staples and consumer discretionary companies were significantly more positive than their replies to tweets of utilities, energy, and telecommunication service companies.

Summary

In this chapter we presented our findings of how the Fortune 500 companies used Twitter for corporate communication purposes and tested whether the companies' Twitter activities were associated with their business performances. We also examined how different business sectors differed in their Twitter activeness. In conclusion, Twitter appears to be a useful social media platform for companies to engage their stakeholders. A company's Twitter activeness is significantly related to its profitability to a certain extent. We will turn to another popular social media platform, Facebook, in the next chapter.

Chapter Five

Facebook

Chapter Overview

This chapter examines how the Fortune 500 companies used Facebook for their corporate nonfinancial communication from January 2009 to December 2013. By analyzing the data collected from Facebook and other sources, we intended to answer the following questions in this chapter:

- How do the Fortune 500 companies use Facebook to communicate with their publics in general?
- How does the public engage with Facebook to communicate with the Fortune 500 companies?
- What is the best way to measure a company's nonfinancial activeness on Facebook?
- Does a company's Facebook nonfinancial activeness affect its business performance from a finance perspective, measured in its monthly stock return?
- Does a company's Facebook nonfinancial activeness affect its business performance from an accounting perspective, measured

in its quarterly total revenue, net income (or loss), earnings per share, profit margin, return on assets, and return on equity?
- Are companies in different business sectors involved in nonfinancial activities on Facebook differently?

Companies' Activities on Facebook

To understand how the Fortune 500 companies used Facebook for their corporate communication, we started collecting the raw data from the Facebook website on November 7, 2013. The data collection process lasted approximately 8 months, ending on July 9, 2014. A total of 407 companies on the Fortune 500 list were found to have corporate accounts on Facebook. Of those 407 corporate accounts, 98 of them were verified by the companies. There were several times in this data collection process, when a company seemed to have multiple accounts, that we had to decide which Facebook account to include in our project. In some of those cases we used the company's corporate website to help make the decision because it provided a hyperlink to a Facebook account: We considered that account to be the company's official Facebook account, and selected it for our research. In some other cases, the company's website did not provide a link to its Facebook account. We then had to make the decision based on a careful examination of the content provided by different accounts. We generally selected the account that reflected the company's core business most closely. Finally, in a scenario where a parent company had no Facebook account but most or all of its subsidiaries had Facebook accounts, we selected the most prominent subsidiary to represent the parent company.

The 407 companies that had Facebook accounts clearly differed from one another in regard to their usage history. Some companies started using Facebook to communicate with the public as early as 2007, but some other companies did not adopt Facebook for corporate communication until 2014.[1] In our database we coded this variable in a way that reflected the number of years each company had been using Facebook. For instance, for a company that started to use Facebook in 2007, this variable would be 8 because the company had been on Facebook for 8 years (from 2007 to 2014).[2]

In addition to how long each company had been using Facebook, we also recorded the total number of "likes" each company earned on its Facebook account. This "like" figure could be regarded a Facebook popularity indicator. On average, each company generated 1,905,479 likes (minimum: 8; maximum: 154,294,821). The top three most popular companies on Facebook during our data collection period appeared to be Facebook, Coca-Cola, and Walt Disney, with 154,294,821; 78,311,991; and 45,804,620 likes, respectively.[3] We performed a correlation analysis with each company's Fortune 500 ranking, Facebook usage history, and total number of likes. The Fortune 500 ranking was not significantly correlated with the other two variables, but *a company's total number of Facebook likes was significantly and positively correlated with its Facebook usage history*.[4] This positive correlation was expected, and easy to understand: The longer a company used Facebook, the more likes it tended to generate.

Companies' Facebook Posts

We used a sampling method similar to that used in the Twitter research to collect Facebook posts from the Fortune 500 companies. For each company that had a Facebook account, one post was randomly selected for each month from January 2009 to December 2013, if the company had at least one post during that month. Since this is a 5-year period with 60 months, we sampled a maximum of 60 Facebook posts for each company.

Once we sampled a Facebook post, we saved its screenshot immediately. There were three metrics associated with each post: (1) the number of "likes," (2) the number of "comments," and (3) the number of "shares." We recorded these three metrics for each post in our database, based on the saved screenshots. Similar to what we did for the Twitter platform, we also recorded the top two comments on each post, if the post generated at least two comments (if a post generated only one comment, we recorded that comment; if it generated no comment, we then considered its comment data to be none [0]).

After we completed this raw data collection process, we ended up with a total of 15,384 Facebook posts. These posts, taken together, could be considered a snapshot of the Fortune 500 companies' communication activities with their publics on Facebook from January 2009 to December 2013. By conducting descriptive statistics, we found that each post generated an average of 960.41 likes (minimum: 0; maximum: 428,818), 87.92 comments (minimum: 0; maximum: 34,466), and 81.01 shares (minimum: 0; maximum: 101,446). We also performed a correlation analysis among these three metrics. We found that they were all significantly and positively correlated.[5] That is to say, *when a post gets popular, more people "like" it, more people "comment" on it, and more people "share" it as well.*

As described earlier, we also collected up to two comments for each post. Using this method, we ended up with a total of 18,008 comments. These comments provided valuable insights about how the public responded to the Fortune 500 companies' posts on Facebook in general. To translate these narrative comments into numerical measures in our database, we used a content analysis approach similar to what we employed in the Twitter research. Again, two graduate students served as the coders (they did not participate in the Twitter coding). They coded each comment into one of the five categories: (1) compliment, (2) complaint, (3) question, (4) self-promotion, and (5) neutral opinion. The definition of each category was the same we used in the Twitter coding. We randomly selected five companies that had Facebook accounts for the coder training. After explaining the definition of each category to the two coders and ensuring that there was no ambiguity, we asked each coder to independently code the sample posts from those five companies. After they completed this training session, their coding results were compared. The overall agreement between the two coders was 91.6%, suggesting a satisfactory level of reliability;[6] divergences between them were later resolved by discussion. The remaining posts were evenly divided into two halves, and each coder coded one of them.

The whole coding process lasted approximately 7 weeks. To illustrate the various kinds of comments the public made on the Fortune 500 companies' posts on Facebook, we selected one example for each category, as follows (all examples are original, and we list them here without correcting any spelling or grammar):

(1) Compliment: "all these outfits are so cute!!"
(2) Complaint: "If you people really feel that way you should stay closed til tomorrow so your employees can do the same."
(3) Question: "When will there be one opening up in Little Rock?"
(4) Self-promotion: "A GREAT FAN PAGE" (to promote a Facebook fan page that was not associated with the company that created the original post)
(5) Neutral opinion: "Amen! God Bless you all'"

In general, the comments on the Fortune 500 companies' Facebook posts appeared to be mostly positive or neutral (31.4% compliment, 10.7% complaint, 6.4% question, 3.8% self-promotion, and 47.7% neutral opinion). The five comment categories were later transformed into three levels in the database according to their valence, with compliment being +1 (positive), complaint being –1 (negative), and all others being 0 (neutral). Since we recorded up to two comments for each post, we averaged their valence scores. Each post's average comment valence ranged from –1 to +1. Finally, we used the following formula to calculate each post's overall comment valence, because each post generated a different number of comments:

Overall comment valence = number of comments × average comment valence

As we pointed out in Chapter 4, this calculation was a rough estimate of each post's overall comment valence. Given that the average number of comments was close to 88 but we only sampled the top two comments for each post, this score might not reflect the overall comment valence 100% precisely.

Companies' Facebook Activities and Their Stock Returns

As described in Chapter 4, a total of 472 companies on the Fortune 500 list appeared to be public, and their monthly stock return information from January 2009 to December 2013 was collected from the database CRSP, using their stock tickers in the search. The monthly return information of the Standard & Poor's 500 Composite Index from January 2009 to December 2013 was also recorded from the same database. To test the relationship between a company's Facebook nonfinancial activeness and its business performance from a finance perspective, we used a hierarchical regression approach similar to what we used in the Twitter research described in Chapter 4. Specifically, we had six variables in our database and we intended to test their relationship in a hierarchical regression analysis. The six variables included:

(1) The monthly stock return of each public company,
(2) The monthly return of the Standard & Poor's 500 Composite Index,
(3) The number of likes (*reputation*) of each Facebook post (it represented a specific month when the post was created on Facebook),
(4) The number of comments (*relationship*) on each Facebook post (it represented a specific month when the post was created on Facebook),
(5) The number of shares (*trust*) of each Facebook post (it represented a specific month when the post was created on Facebook), and
(6) The overall comment valence score (*credibility, confidence*) of each Facebook post (it represented a specific month when the post was created on Facebook).

The purpose of this hierarchical regression analysis was to examine whether the numbers of likes, comments, and shares and the overall comment valence as nonfinancial variables would significantly influence a company's monthly stock return, independent of the effect of the Standard & Poor's 500 Composite Index's monthly return on the company's stock price change. We standardized all six variables before putting them into the following two regression equations, because several of them were significantly correlated with each other.[7] By comparing equation 1 and equation 2, we could determine whether any of the four Facebook nonfinancial indicators (the number of likes, the number of comments, the number of shares, and the overall comment valence) had any significant and independent impact on a company's monthly stock return.

Equation 1: $y = a + b_1x_1 + e$, *and*
Equation 2: $y = a + b_1x_1 + b_2x_2 + b_3x_3 + b_4x_4 + b_5x_5 + e$, where

y = the monthly stock return of each public company
a = constant
x_1 = the monthly return of the Standard & Poor's 500 Composite Index
x_2 = the number of likes of each Facebook post
x_3 = the number of comments on each Facebook post
x_4 = the number of shares of each Facebook post
x_5 = the overall reply valence score of each Facebook post
b_1 to b_5 = regression coefficients
e = residue

The regression analysis obtained significant results. First, a company's monthly stock return was significantly affected by the Standard & Poor's 500 Composite Index's monthly return.[8] This effect should be expected. Second, more importantly, *the number of comments had a significant and positive impact on a company's monthly stock return*, and this effect was independent of the influence of the Standard and Poor's 500 Composite Index.[9] In other words, the more comments a company saw on its Facebook account for a specific month, the more likely it would have a higher level of stock return for that month, no matter how the Standard and Poor's 500 Composite Index changed. Finally, the numbers of likes and shares and the overall comment valence had no significant effect on the monthly stock return.[10]

Companies' Facebook Activities and Their Profitability

Besides testing the relationship between a company's Facebook activeness and its monthly stock return, we also wanted to investigate whether a company's Facebook nonfinancial activeness would be significantly associated with its profitability. As seen in Chapter 4, we recorded several profitability measures from the database Compustat for each public company on the Fortune 500 list from January 2009 to December 2013, including its total revenue, net income (or loss), earnings per share, total assets, and total equity, all on a quarterly basis. Using those figures, we also calculated each company's quarterly profit margin, return on assets, and return on equity. In summary, we used six indicators to measure a company's profitability, including its quarterly (1) total revenue, (2) net income (or loss), (3) earnings per share, (4) profit margin, (5) return on assets, and (6) return on equity.

Each company's quarterly Facebook activeness was not recorded from an existing database, but calculated by us. Specifically, for those companies that ended their fiscal year in December, we added their numbers of likes, comments, and shares and overall comment valence scores in January, February, and March to represent their first quarter's numbers of likes, comments, and shares and overall comment valence. We did similar calculations for each company's second quarter, third quarter, and fourth quarter. For those companies that did not end their fiscal year in December, we adjusted our calculations accordingly. For example, for a company that ended its fiscal year in February, the first quarter's number of likes would be a sum of March's, April's, and May's.

Finally, we recorded the business sector to which each company belonged, from the database Compustat, according to the Global Industry Classification Standard. Nine dummy variables were created for business sectors and three additional dummy variables were created for seasonality, so that we could control the effect of business sector and seasonality on a company's profitability in our data analyses.[11] Specifically, we conducted six separate hierarchical regression analyses to examine whether a company's quarterly numbers of likes, comments, and shares and overall comment valence could significantly affect its quarterly total revenue, net income (or loss), earnings per share, profit margin, return on assets, and return on equity, independent of the effect of business sector and seasonality. We standardized all the variables in the analyses except for the dummy variables, in order to avoid any potential collinearity problem.[12] The specific regression equations that we tested were as follows:

Equation 1: $y = a + b_1x_1 + \ldots + b_3x_3 + b_4x_4 + \ldots + b_{12}x_{12} + e$, *and*
Equation 2: $y = a + b_1x_1 + \ldots + b_3x_3 + b_4x_4 + \ldots + b_{12}x_{12} + b_{13}x_{13} + b_{14}x_{14} + b_{15}x_{15} + b_{16}x_{16} + e$, where

y = each public company's quarterly total revenue, net income (or loss), earnings per share, profit margin, return on assets, or return on equity
a = constant
x_1 to x_3 = three dummy variables for seasonality
x_4 to x_{12} = nine dummy variables for business sector
x_{13} = the quarterly number of likes
x_{14} = the quarterly number of comments
x_{15} = the quarterly number of shares
x_{16} = the quarterly overall comment valence score
b_1 to b_{16} = regression coefficients
e = residue

We found several significant results based on these hierarchical regression analyses. First, the *quarterly number of likes a company generated on Facebook was significantly and positively related to its quarterly total revenue, net income, and return on assets.*[13] Second, the *quarterly number of comments was found to have a significant and positive impact on the quarterly total revenue.*[14] Third, the *quarterly number of shares had a significant but negative effect on the quarterly total revenue.*[15] Finally, the *quarterly overall comment valence had significant negative effects on both the quarterly total revenue and profit margin.*[16]

The first two findings seemed to be easy to interpret. When more consumers "like" a company, it helps the company generate a higher level of revenue, net income, and return on assets, probably because liking is a strong indicator of third-party endorsement. However, the third finding and the fourth finding were somewhat surprising. Why would the number of shares and the overall comment valence have *negative* effects on the total revenue and profit margin? To answer this question, we tested a possible explanation that these findings might be caused by a few outliers in the database. Thus, we deleted those outliers and conducted the hierarchical regression analyses again. The significant effects of the overall comment valence on the total revenue and profit margin, as well as the significant effect of the number of shares on the quarterly total revenue that we discovered earlier, disappeared, confirming that a few outliers might have created those artificial effects.[17] Another possible explanation is related to the limitation of our data. In the raw data collection process, we recorded only the top two comments for each post. There was a chance that the valence of the top two comments significantly differed from the other comments. Therefore, the overall comment valence that we calculated might not be 100% accurate.

Finally, we conducted several independent t-tests to examine whether a company's Facebook activeness during a profitable quarter differed from that during a nonprofitable quarter. We used the same profitability variable that we created in the Twitter research to differentiate two types of quarters: profitable quarters with positive net incomes and nonprofitable quarters with negative net losses. The numbers of likes, comments, and shares and the average comment valence were compared between these two groups of quarters. Two significant differences were detected. First, the *companies generated significantly more likes in profitable quarters (approximately 3,196 likes on average) than in nonprofitable quarters (approximately 1,223 likes on average).*[18] Second, the *average comment valence for a profitable quarter (approximately 0.38 on a scale from –1 to +1) was significantly more positive than that for a nonprofitable quarter (approximately 0.18 on a scale from –1 to +1).*[19] No significant difference was found between the two sorts of quarters in terms of the numbers of comments and shares.[20]

Companies' Facebook Activities Across Business Sectors

The last data analysis that we conducted on Facebook was a series of one-way ANOVA tests. The objective of these tests was to compare the quarterly numbers of likes, comments, and shares and the average comment valence across 10 different business sectors, so as to understand which business sectors tended to be more or less active on Facebook. As described earlier, the categorization of those 10 business sectors was based on the Global Industry Classification Standard, and included energy, materials, industrials, and seven others.

There were several significant findings. First, *different business sectors generated different numbers of likes for each quarter*. The consumer discretionary sector was the clear "winner" in this aspect. It generated significantly more likes on average (approximately 6,800.28) than all other sectors except for information technology and telecommunication services (ranged from 63.85 to 2,796.04), based on an ANOVA test and a post hoc analysis.[21]

Second, the *information technology sector was the most successful in generating user comments on Facebook*. According to an ANOVA test and a post hoc analysis, the average number of comments generated by the information technology sector (approximately 624.63) was significantly higher than all other sectors except for consumer discretionary and telecommunication services (ranged from 6.89 to 304.99).[22]

Next, we compared each business sector with regard to the number of shares. There was no single "winner" in this case. The only significant differences were that the *Facebook posts from the industrial sector were shared significantly fewer times (approximately 76.80) than those from the information technology sector (approximately 517.42) and the consumer discretionary sector (approximately 491.80)*.[23]

Finally, we examined whether there was any significant difference across various sectors regarding the average comment valence. There was a clear "loser" in this regard: telecommunication services. An ANOVA test and a post hoc analysis showed that the *average comment valence on the Facebook posts by the telecommunication services sector (approximately –0.33) was significantly more negative than all other sectors (ranged from 0.13 to 0.47)*.[24]

Limitations of Our Data

Before we close this chapter, we would like to address a couple of limitations associated with our data in this Facebook project. The biggest weakness of our data is that they were collected over a long period (about 8 months), potentially resulting in a certain level of measure error. Specifically, we collected the raw data from Facebook for each company based on its placement on the 2013 Fortune 500 list. This process was quite lengthy because we needed to randomly sample one post and its top two comments for each company and for each month in a 5-year time frame (from January 2009 to December 2013). Once we sampled a post, we saved a screenshot of it on a computer and then recorded the numerical metrics associated with it in our database, including its numbers of likes, comments, and shares. When we finished the whole raw data collection process, we ended with more than 15,000 posts. Clearly, this procedure could not be finished within a single day or a few days. Indeed, it took months to complete. The chances are that when we finished the data collection for one company and moved on to another, the data for the first company might have changed a bit on Facebook. For example, a company's post might have 100 likes when we collected the data, but the number could change to 101 the next day. Although it is unlikely that a Facebook user would go back (especially, way back) and "like," "comment" on, or "share" a historical post, we have to admit that this possibility does exist. This appears to be a unique feature attached to social media. From a data analysis viewpoint, it is unlikely that a minor change in those metrics will significantly impact our overall findings, because we have thousands of cases in our database, but we want our readers to be aware of this specific shortcoming of the data.

Second, we used a sampling method to collect user comments for this Facebook research that is similar to what we did for Twitter. In particular, we sampled the top two comments on each post, and we estimated the post's average comment valence by coding those two comments as positive, neutral, or negative. Given that each Facebook post in our database generated an average of 88 comments, this coding might not reflect the true comment valence perfectly. We chose to use this sampling method for two reasons. First, we wanted to be consistent with the method that we used in the Twitter data gathering process. Second, such a sampling method was a systematic and fair choice with regard to feasibility, because simple random sampling would be extremely difficult to implement (for example, in the most extreme case, a Facebook post generated 34,466 comments). However, we want the readers to be aware of this sampling's weakness, even though it is a reasonable choice.

Synopsis of Research Findings

This chapter focused on how the Fortune 500 companies used Facebook for their corporate communication from 2009 to 2013. Several significant results were detected in our data analyses. The key findings can be synopsized as follows:

- In general, Facebook seemed to be a popular platform for corporate communication, because 407 companies on the Fortune 500 list were found to have Facebook accounts during our data collection period.
- Some companies started using Facebook to communicate with the public as early as 2007, but some others did not adopt Facebook until 2014. The longer a company used Facebook, the more "likes" it tended to generate from the public (*reputation*).
- Compared to Twitter, the public showed more enthusiastic involvement in their conversations with the Fortune 500 companies on Facebook. On average, a company's Facebook post generated about 960 likes (*reputation*), 88 comments (*relationship, credibility, confidence*), and 81 shares (*trust*). By comparison, a company's Twitter tweet generated an average of about 4 favorites, 8 retweets, and less than 1 comment.
- The numbers of likes, comments, and shares generated by a Facebook post were positively correlated, which means that when a company's post gets attention from the public, they tend to "like" it, "comment" on it, and also "share" it with their friends (*relationship, reputation, trust, credibility, confidence*).
- When the public commented on the Fortune 500 companies' Facebook posts, most often the comments were either positive or neutral (*credibility, confidence*).

- The average number of comments (*relationship*) a company generated on Facebook each month had a significant and positive impact on the company's monthly stock return, and this effect was independent of the Standard & Poor's 500 Composite Index's influence on the company's stock price change.
- The number of likes (*reputation*) that a company earns on Facebook can "predict" its profitability. Specifically, we found in our data analyses that the average number of likes a company received each quarter on Facebook was significantly and positively associated with its quarterly total revenue, net income (or loss), and return on assets.
- We also found that the average number of comments (*relationship*) a company generated each quarter on Facebook was significantly and positively related to its quarterly total revenue.
- The quarterly number of shares (*trust*) had a significant but negative effect on the quarterly total revenue in our data analyses. Moreover, the comment valence (*credibility, confidence*) was found to be significantly but negatively associated with the quarterly total revenue and profit margin. These two findings need to be interpreted with caution, because these effects disappeared when we excluded a few outliers in the data analyses, which suggested that they might be artificial.
- A company tended to generate more likes (*reputation*) on Facebook during a profitable quarter than during a nonprofitable quarter. The comments on its posts (*credibility, confidence*) during a profitable quarter were also significantly more positive than those during a nonprofitable quarter.
- Different business sectors appeared to differ with regard to their Facebook nonfinancial activeness. The consumer discretionary sector was the "winner" in generating likes on Facebook, while the information technology sector seemed to be the most successful in generating comments. Regarding the number of shares, Facebook posts from the industrial sector were shared significantly fewer times than those from the information technology and consumer discretionary sectors. Finally, the telecommunications services

sector was found to be the "loser" regarding the average comment valence. The comments on Facebook posts from companies in the telecommunication services sector were significantly more negative than those from all other sectors.

Summary

The focus of this chapter is Facebook. We examined how active the Fortune 500 companies were on Facebook and how the public engaged in dialogues with the companies in general. We also tested the relationships between the companies' Facebook nonfinancial activeness and their business performance, from both a finance perspective and an accounting perspective. The research findings, in general, showed a significant relationship between a company's Facebook activeness and its business performance. In the next chapter, we will discuss the third social media platform included in our research: YouTube.

Chapter Six

YouTube

Chapter Overview

This chapter is focused on the video messages that the Fortune 500 companies delivered to their publics via YouTube from January 2009 to December 2013. We collected the raw data from the companies' YouTube accounts and other sources. By analyzing those data, we aimed to answer the following questions:

- What kinds of videos do the Fortune 500 companies put on YouTube?
- How does the public respond to the Fortune 500 companies' videos on YouTube, in general?
- What is the best way to measure a company's activeness on YouTube?
- Does a company's YouTube nonfinancial activeness influence its business performance from a finance perspective, measured in its monthly stock return?
- Does a company's YouTube nonfinancial activeness impact its business performance from an accounting perspective, measured

in its quarterly total revenue, net income (or loss), earnings per share, profit margin, return on assets, and return on equity?

- Are there significant differences with regard to different business sectors' nonfinancial activeness on YouTube?

Companies' Activities on YouTube

The data collection for the YouTube platform started at about the same time as the Twitter platform, but it lasted longer. Specifically, we started collecting the raw data from the YouTube website on January 26, 2014, and completed the whole data collection process on July 17, 2014. A total of 392 companies on the Fortune 500 list were found to have corporate accounts on YouTube. Among those 392 accounts, 75 were verified by the companies. The YouTube website provided information on when each account was set up, and we recorded that information. There was wide variation in this regard: A few companies made their debuts on YouTube as early as 2005, but some late adopters did not establish their presence on YouTube until 2013.[1] In our database we coded this variable numerically to show how long a company had been on YouTube, in a manner similar to what we did in the Facebook research. For example, for a company that established its YouTube account in 2005, this variable would be 9 (counting from 2005 to 2013).[2]

We also recorded three key metrics associated with each company on its YouTube account, including (1) the number of videos, (2) the number of subscribers, and (3) the number of views. The number of videos could reflect how much effort a company had made to communicate with its publics on YouTube, while the numbers of subscribers and views could suggest how enthusiastic the public was about the company's communication messages. We found that, on average, each company uploaded 385 videos on YouTube (minimum: 1; maximum: 57,195). Each company had an average of 14,238 subscribers (minimum: 0; maximum: 1,812,366) and 9,314,198 views (minimum: 9; maximum: 1,062,088,358). The most active companies on YouTube during our data collection period were CBS, Intel, and Cisco Systems, producing 57,195; 5,102; and 4,101 videos, respectively. The top three most popular companies on YouTube appeared to be Apple (with 1,812,366 subscribers), CBS (with 541,650 subscribers), and Coca-Cola (with 223,417 subscribers). Furthermore, CBS, Google, and Coca-Cola were the most successful in generating video views, having 1,062,088,358; 812,018,605; and 233,157,587 views, respectively.[3]

We conducted correlation analyses with a company's total numbers of videos, subscribers (*credibility*), and video views, its YouTube usage history, and its Fortune 500 ranking. Several significant effects were detected. First, a company's total number of subscribers was significantly and positively correlated with its YouTube history. That is to say, the longer a company uses YouTube for corporate communication, the more subscribers it attracts.[4] Second, a company's Fortune 500 ranking was significantly correlated with its YouTube usage history and total number of subscribers. Specifically, the higher a company ranks on the Fortune 500 list, the longer it tends to have been on YouTube and the more subscribers it is likely to have as well.[5] Finally, a company's total numbers of videos, subscribers, and video views were all significantly and positively correlated, which meant that the more YouTube videos a company produced, the more subscribers and views it generated.[6]

Companies' YouTube Videos

To be consistent with the methodology described in the Twitter and the Facebook chapters, we again did a systematic sampling in this YouTube project. For each company that had a YouTube account, one video was randomly selected for each month from January 2009 to December 2013, when the company had at least one video listed for that month. Using this sampling method, a maximum of 60 videos would be sampled for each company.

Once we sampled a video, we immediately saved its web address in a Word document. We also recorded six metrics associated with the video in our database, including (1) the number of "views," (2) the number of "likes" (*reputation*), (3) the number of "dislikes" (*reputation*), (4) the number of "comments" (*relationship*), (5) the number of "shares" (*trust*), and (6) the video's length (counted in seconds).

Moreover, we collected the top two comments on each sampled video, as we did in the Twitter and Facebook research (if a video had only one comment, we recorded that comment; if it had no comment, we then considered its comment data to be none [0]). To ensure that we could revisit those comments in a later content analysis, we saved a screenshot of each video. During this data collection process, we noticed that YouTube allowed a company to disable the comment function for its videos, which could indicate distrust between a company and its public or other relationship problems. Thus, we recorded whether the company had disabled the comment function for each sampled video.[7]

After we repeated this sampling procedure for each company, we had collected a total of 7,885 videos. These videos differed significantly in length. The shortest one was just 4 seconds, while the longest was 3 hours 26 minutes 42 seconds. On average, each video was about 3 minutes 31 seconds long. As reflected in the data, the public showed a high level of involvement with the videos; each video was viewed an average of 33,989 times (minimum: 1; maximum: 30,960,854). The viewers, in general, more often expressed positive reputation ("likes" average: approximately 87; minimum: 0; maximum: 55,963) than negative reputation ("dislikes" average: approximately 9; minimum: 0; maximum: 4,763) in regard to each video. On average, a video generated about 13 comments (minimum: 0; maximum: 9,699) and was shared about 40 times (minimum: 0; maximum: 38,813).

Using the method described earlier, we also collected a total of 5,168 video comments. To understand how the public responded to the companies' videos in these comments, we implemented a content analysis approach similar to what we used in the Twitter and Facebook research. In particular, we employed four graduate students as coders, none of whom had participated in the Twitter or Facebook coding. We held a training session for the coders and explained the coding procedure to them in detail. They were instructed to code each comment into one of the five categories: (1) compliment, (2) complaint, (3) question, (4) self-promotion, and (5) neutral opinion.[8] To check the reliability of the coders' coding, we randomly selected from our database five companies that had YouTube accounts. The four coders were asked to independently code all the comments on those five companies' YouTube videos. The average agreement of their coding was 92.3%, suggesting a high level of reliability. The discrepancies in the coding were later resolved by a discussion among the coders. After the training session, we divided the remaining videos into four parts, and each coder coded one part.[9] The whole coding procedure lasted approximately 6 weeks.

For illustration purposes, we list one example for each category as follows (all examples are original, without corrections of spelling or grammar):

(1) Compliment: "oh my god so beautiful:)"
(2) Complaint: "I just bought a head of lettuce for $2.99 and it was full of dirt, disgusting"
(3) Question: "Whats this called?????"
(4) Self-promotion: "I have a gary fisher advance for sale for $140 on craigslist im from richmond type in craigslist Gary Fisher Advance and email me from craigslist or send me a message on youtube"
(5) Neutral opinion: "I saw me 0–0"

Overall speaking, the comments looked mostly positive or neutral (32.1% compliment, 9.7% complaint, 6.3% question, 12.6% self-promotion, and 39.3% neutral opinion). As we did in the Twitter and Facebook research, we recoded these categories into three levels in the database based on their valence, with compliment being +1 (positive), complaint being –1 (negative), and everything else being 0 (neutral). Since each sampled video had up to two comments in our database, we then averaged the valence scores of the comments. As expected, the range of each video's average comment valence score was from –1 to +1. To calculate its overall comment valence score, we used the following formula:

Overall comment valence = number of comments × average comment valence

Although the viewers' attitudes in the comments appeared to be mostly non-negative, we did find that the comment function was disabled for 15.7% of the videos. To examine whether the viewers would react differently when a company disabled the comment function of its videos, we conducted a series of independent t-tests. It was found that *people gave significantly more "dislikes" to a video when its comment function was disabled (approximately 21 dislikes) than it was not (approximately 7 dislikes).*[10]

Companies' YouTube Activities and Their Stock Returns

To test the effect of a company's YouTube activeness on its business performance from a finance perspective, we adopted an approach similar to that described in previous chapters. The companies that we included in this analysis were those 472 public companies on the Fortune 500 list, since their monthly stock return information was recorded via the database CRSP. In particular, we set up nine variables in our database, reflecting each company's YouTube activeness and its monthly stock return, as well as the Standard & Poor's 500 Composite Index's monthly return. The variables are as follows:

(1) The monthly stock return of each public company,
(2) The monthly return of the Standard & Poor's 500 Composite Index,
(3) The number of views of each YouTube video (it represented a specific month when the video was created on YouTube),
(4) The number of likes (*reputation*) of each YouTube video (it represented a specific month when the video was created on YouTube),
(5) The number of dislikes (*reputation*) of each YouTube video (it represented a specific month when the video was created on YouTube),
(6) The number of comments (*relationship*) on each YouTube video (it represented a specific month when the video was created on YouTube),
(7) The number of shares (*trust*) of each YouTube video (it represented a specific month when the video was created on YouTube),
(8) The length of each YouTube video (it represented a specific month when the video was created on YouTube), and

(9) The overall comment valence score (*credibility, confidence*) of each YouTube video (it represented a specific month when the video was created on YouTube).

We entered these nine variables into two regression equations to test if any of the YouTube activeness indicators (the numbers of video views, video likes, video dislikes, video comments, and video shares, the video length, and the overall comment valence) had a significant impact on a company's monthly stock return, independent of the effect of the Standard & Poor's 500 Composite Index. Since several of these variables were significantly correlated with each other, we standardized their values before entering them in the regression equations.[11]

Equation 1: $y = a + b_1x_1 + e$, *and*
Equation 2: $y = a + b_1x_1 + b_2x_2 + b_3x_3 + b_4x_4 + b_5x_5 + b_6x_6 + b_7x_7 + b_8x_8 + e$, where

y = the monthly stock return of each public company
a = constant
x_1 = the monthly return of the Standard & Poor's 500 Composite Index
x_2 = the number of views of each YouTube video
x_3 = the number of likes of each YouTube video
x_4 = the number of dislikes of each YouTube video
x_5 = the number of comments on each YouTube video
x_6 = the number of shares of each YouTube video
x_7 = the length of each YouTube video
x_8 = the overall comment valence score of each YouTube video
b_1 to b_8 = regression coefficients
e = residue

The analysis results obtained significance for equation 1, but insignificance of the incremental change from equation 1 to equation 2. That is to say, the Standard & Poor's 500 Composite Index's monthly return had a significant influence on a company's monthly stock return.[12] However, *none of the YouTube activeness indicators showed a significant effect on a company's monthly stock return.*[13]

Companies' YouTube Activities and Their Profitability

We also examined whether a company's YouTube nonfinancial activeness was significantly associated with its business performance from an accounting perspective. Again, we included the 472 public companies on the Fortune 500 list in this data analysis because their financial data were assessable in public databases. As described in Chapters 4 and 5, we recorded five quarterly financial indicators for each public company from the database Compustat, including the total revenue, net income (or loss), earnings per share, total assets, and total equity. Based on such information, we calculated three other indicators, including the profit margin, return on assets, and return on equity. In total, we used six quarterly measures to reflect a company's profitability: (1) total revenue, (2) net income (or loss), (3) earnings per share, (4) profit margin, (5) return on assets, and (6) return on equity.

To reflect a company's quarterly YouTube activeness, we used seven measures: (1) the number of video views, (2) the number of video likes, (3) the number of video dislikes, (4) the number of video comments, (5) the number of video shares, (6) the video length, and (7) the overall video comment valence. These measures would indicate not only how much of an effort a company had made on YouTube to communicate with its publics, but also how well the communication messages were received and interpreted by the public. We transformed each company's monthly YouTube activeness figures into quarterly figures by adding the data of 3 months for each quarter. In this calculation process, we adjusted the formula for those companies that did not end their fiscal year in December, as detailed in the earlier chapters.

Before entering these financial indicators and YouTube activeness indicators into the regression equations for statistical tests, we standardized their values because several of them were significantly correlated with each other.[14] We also included the dummy variables of business sector and seasonality in the regression equations because we wanted to control their effects.[15] In summary, we conducted six hierarchical regression analyses, testing the effect of a company's quarterly YouTube activeness on its quarterly total revenue, net income (or loss), earnings per share, profit margin, return on assets, and return on equity, respectively. The specific regression equations were as follows:

Equation 1: $y = a + b_1x_1 + \ldots + b_3x_3 + b_4x_4 + \ldots + b_{12}x_{12} + e$, and
Equation 2: $y = a + b_1x_1 + \ldots + b_3x_3 + b_4x_4 + \ldots + b_{12}x_{12} + b_{13}x_{13} + b_{14}x_{14} + b_{15}x_{15} + b_{16}x_{16} + b_{17}x_{17} + b_{18}x_{18} + b_{19}x_{19} + e$, where

y = each public company's quarterly total revenue, net income (or loss), earnings per share, profit margin, return on assets, or return on equity
a = constant
x_1 to x_3 = three dummy variables for seasonality
x_4 to x_{12} = nine dummy variables for business sector
x_{13} = the quarterly number of YouTube video views
x_{14} = the quarterly number of YouTube video likes
x_{15} = the quarterly number of YouTube video dislikes
x_{16} = the quarterly number of YouTube video comments
x_{17} = the quarterly number of YouTube video shares
x_{18} = the quarterly YouTube video length
x_{19} = the quarterly overall YouTube video comment valence score
b_1 to b_{19} = regression coefficients
e = residue

Several significant results were obtained. First, the number of YouTube video likes a company received each quarter had a significant and positive impact on its quarterly total revenue, net income (or loss), and earnings per share.[16] Second, the number of YouTube video dislikes a company got each quarter also had a significant and positive effect on its quarterly total revenue and net income (or loss).[17] Third, the number of video shares had a significant effect on a company's quarterly total revenue and net income (or loss), but the effect was negative.[18] No other significant result was found in the regression analyses.[19]

The first finding was easy to comprehend. It was consistent with what we discovered in the Facebook research. *When the public gave more "likes" to a company's YouTube videos, it was a powerful signal of third-party endorsement.* The level of such endorsement positively reflected the company's ability to make profits. The second finding looked both interesting and counterintuitive. *The number of video "dislikes" was in fact* positively *related to a company's profitability.* This finding appeared to be robust because the effects remained significant after we reconducted the analyses excluding a few outliers in the database.[20] It may be interpreted from two angles. From a statistical viewpoint, the relationship between a company's YouTube activeness and its profitability might not be causal, although we treated them in a causal fashion in the regression analyses. In other words, it is possible that a company made a big profit by engaging in unethical or unresponsive behaviors, leading the public to "dislike" its YouTube videos. From a theoretical viewpoint, we might explain this finding by referencing the well known proverb "any publicity is good publicity."[21] When people "dislike" something, it shows that they are paying attention to that thing. It is possible that the valence of the public's attention diminishes over time, but the attention itself helps a company to gain revenues and net incomes.

The third finding also appeared to be counterintuitive. To examine whether the results were caused by a few outliers in the database, we reconducted the regression analyses excluding those outliers. *The negative effect of the number of video shares on the net income (or loss) disappeared, but effect on the total revenue remained significant.*[22] We suspected that this finding was caused by a limitation of our data which arose because we recorded a video's "share" figure from its "statistics" report on the YouTube website. However, some companies chose not to provide these reports to the public, so the numbers of their video shares were missing (though the numbers of views, likes, dislikes, and comments on their videos were recorded). Therefore, the data on video shares in our database was incomplete compared to the data on video views, likes, dislikes, and comments, and they might be skewed in some way and generate misleading results.

To explore whether a company's YouTube nonfinancial activeness in a profitable quarter would differ from that in a nonprofitable quarter, we conducted a series of independent t-tests. As described in earlier chapters, we classified each quarter as a profitable quarter or nonprofitable quarter for each company based on the company's reported net income (or loss). We compared the two types of quarters in terms of the numbers of video views, video likes, video dislikes, video comments, and video shares, the video lengths, and the average video comment valence. It was found that a company's videos were viewed significantly more times and liked and disliked more when it was profitable (*reputation*) (video views: approximately 83,220; video likes: approximately 195; video dislikes: approximately 20) than when it was not (video views: approximately 22,421; video likes: approximately 51; video dislikes: approximately 5). The public also tended to provide more comments (*relationship*) on a company's videos when it was profitable (approximately 31 comments) than when it was not (approximately 16 comments), and the comments were more positive (*credibility, confidence*) as well (average comment valence in profitable quarters = 0.17; average comment valence in nonprofitable quarters = 0.12).[23] However, the number of video shares did not significantly differ between the two types of quarters. The video length did not significantly differ, either.

Companies' YouTube Activities Across Business Sectors

Our last analysis in this YouTube project was to examine how various business sectors differed in their YouTube nonfinancial activeness. We conducted seven one-way ANOVA analyses, testing the differences among 10 business sectors with regard to the quarterly numbers of YouTube video views, likes, dislikes, comments, and shares, the video lengths, and the overall video comment valence. The 10 business sectors that we included in these analyses were the same as described in earlier chapters.

Based on the ANOVA test results, the information technology sector appeared to be a "winner" in using YouTube for corporate communication. The *YouTube videos produced by the information technology sector each quarter were significantly longer* (approximately 641 seconds) than the videos produced by any other sector (ranged from approximately 297 seconds to 455 seconds).[24] The videos from the information technology sector also *generated significantly more likes* (approximately 773 likes) and dislikes (approximately 80 dislikes) than the videos from all other sectors (approximately 5 to 160 likes and 1 to 4 dislikes).[25,26] In addition, the information technology sector was more successful *in generating video views* (approximately 248,757 views) and *video comments* (approximately 115 comments) than most other sectors (approximately 3,199 to 81,782 views and 1 to 27 comments).[27,28]

In regard to the video comment valence, the consumer discretionary sector was the most successful. The viewers' *comments on the videos from the consumer discretionary sector were significantly more positive* (valence score was approximately 0.31) *than those on the videos from all other sectors* (ranged from approximately –0.01 to 0.19), *except for industrials.*[29] Finally, there was not much difference among the 10 business sectors regarding the number of video shares. The only significant difference was that the number of video shares (*trust*) generated by the information technology sector was significantly higher than that of consumer discretionary and industrials.[30]

Limitations of Our Data

There are certain limitations associated with our data that need to be addressed. First, as described earlier, the number of video shares in the database might not be as reliable as the numbers of video views, video likes, video dislikes, and video comments, because some companies did not release their video shares information to the public. Second, there were several cases where we sampled a video, but later we found that the company had removed the video from its YouTube account (the YouTube website displayed the message "This video is unavailable"). In these cases, we regarded the video's data as missing. However, this might introduce a certain level of bias to the data. Third, we found that some sampled videos were not publicly accessible because the company chose to make them available to certain viewers only (the YouTube website displayed the message "This video is private"). In such cases, we had to regard the data of those videos as missing, though they were not. We suspect that this might introduce some biases to our dataset.

Furthermore, we noticed that the metrics associated with a YouTube video such as the number of video views tended to change more rapidly than the metrics associated with a Twitter tweet or a Facebook post. People may consider YouTube to be a large video content "library." It is exciting to check out the most recent videos, but it is also fun to view historical videos. Logically, it seems less likely that a person would read an outdated Twitter tweet or Facebook post, and more likely for him or her to view an old YouTube video. Since the raw data collection for YouTube lasted for several months, it was possible that the data we recorded from the saved screenshots were somewhat different from the real-time data. As discussed in Chapter 5, this shortcoming is caused by the nature of social media, and it seems unavoidable. Theoretically speaking, even if we had recorded the data for the Fortune 500 companies within one day, it still is possible that some data would not be real-time because social media metrics can change in a second.

Synopsis of Research Findings

To synopsize the research findings presented in this chapter, we first looked at how the Fortune 500 companies utilized YouTube as a communication platform in general. Then we examined how the public responded to the companies' YouTube videos by liking, disliking, sharing, and commenting. We also tested the relationship between a company's YouTube activeness and its business performance. Finally, we analyzed the data to see how various business sectors differed in their usage of YouTube. The key findings are:

- YouTube is a popular social media platform for the Fortune 500 companies to communicate with their publics. A total of 392 companies on the Fortune 500 list were found to have corporate accounts on YouTube during our data collection period.
- There were certain differences among the Fortune 500 companies regarding their YouTube usage history. Some companies started to deliver videos to their publics via YouTube as early as 2005, and some other companies did not start using YouTube until 2013.
- Overall, the Fortune 500 companies were active YouTube users. On average, each company generated 385 videos, 14,238 subscribers, and 9,314,198 video views.

- A company's YouTube usage history and its number of subscribers (*credibility*) were significantly correlated with its Fortune 500 ranking. *The higher it ranked, the earlier it had begun to use YouTube and the more subscribers it was likely to have.*
- On average, each YouTube video created by one of the Fortune 500 companies was about 3.5 minutes long. It generated approximately 33,989 views, 87 likes, 9 dislikes, 13 comments, and 40 shares.
- Similar to Twitter and Facebook, when the public made comments on the videos on YouTube, they tended to be mostly positive or neutral (*confidence*).
- When a company disabled the comment function for its videos, the public expressed significantly more "dislikes" (*reputation*).
- We tested the relationship between a company's YouTube activeness and its monthly stock return. *No significant effect was found.*
- The number of video likes (*reputation*) a company received on YouTube each quarter was found to have a *significant and positive effect on its quarterly total revenue, net income (or loss), and earnings per share.*
- The number of video dislikes (*reputation*) a company generated on YouTube each quarter also had a significant and positive effect on its quarterly total revenue and net income (or loss). This finding is counterintuitive, and there are several possible explanations.
- The number of YouTube video shares (*trust*) was found to have a significant but negative effect on a company's quarterly total revenue and net income (or loss). We suspect that this finding was caused by certain limitations of our data, and we suggest that our readers interpret this result with caution.
- When a company made a profit in a quarter, its YouTube videos tended to generate significantly more views, likes, dislikes, and comments (*relationship, reputation, credibility*). The viewers' comments were also likely to be more positive (*confidence*).
- Different business sectors differed in their YouTube activeness. The videos produced by the information technology sector were significantly longer than the videos produced by other sectors. The information technology sector also appeared to be the most

successful in generating video views, likes, dislikes, and comments (*relationship, reputation, credibility*). Regarding the video comment valence (*confidence*), the viewers' comments on the videos produced by the consumer discretionary sector were the most positive.

Summary

In this chapter we discussed how the Fortune 500 companies used YouTube for their corporate communication and how using YouTube was connected to the companies' business performances. Similar to Twitter and Facebook, YouTube also shows great value in helping companies to engage their stakeholders and improve business profitability. In the next chapter we will turn our focus to the last social media platform, Google+.

Chapter Seven

Google+

Chapter Overview

This chapter is centered on Google+, the last social media platform from which we collected data. Based on an examination of the Fortune 500 companies' activities on Google+ and their business performances, we intended to answer the following questions:

- What kinds of posts do the Fortune 500 companies create on Google+?
- How does the public respond to the Fortune 500 companies' Google+ posts in general?
- What is the best way to measure a company's nonfinancial activeness on Google+?
- Does a company's Google+ nonfinancial activeness affect its business performance from a finance perspective, measured in its monthly stock return?
- Does a company's Google+ nonfinancial activeness influence its business performance from an accounting perspective, measured

in its quarterly total revenue, net income (or loss), earnings per share, profit margin, return on assets, and return on equity?

- How do various business sectors differ in their Google+ non-financial activeness?

Companies' Activities on Google+

To be consistent with the research described in the preceding three chapters (Twitter, Facebook, and YouTube), we adopted a similar research methodology in this Google+ project with regard to both data collection and data analyses. The raw data collection from the Google+ website lasted approximately 1.5 months, starting on May 12, 2014, and ending on June 30, 2014. During this data collection period, we found that a total of 367 companies on the Fortune 500 list had corporate accounts on Google+. Among the 367 corporate accounts, 150 of them were verified by the companies. For those companies that had a Google+ account, we recorded three metrics from their homepages: (1) the number of followers, (2) the number of views, and (3) the number of people in the company's circles. The first two metrics would generally indicate how popular a company was on Google+, and the third one would suggest how actively the company sought to establish a relationship with its publics. It was found that, on average, each company had approximately 104,589 followers (minimum: 1; maximum: 4,180,541), 6,413,402 views (minimum: 0; maximum: 1,487,476,221), and 276 people in its circles (minimum: 1; maximum: 7,678). The top three most popular companies on Google+ during our data collection period appeared to be Google itself, Ford, and Starbucks, having 4,180,541; 3,122,545; and 2,602,754 followers, respectively. Furthermore, Google, eBay, and Yahoo! were the most successful in generating views, having 1,487,476,221; 156,649,805; and 60,554,304 views, respectively.[1]

We conducted correlation analyses among the three metrics, as well as a company's ranking on the Fortune 500 list. It was found that a company's total number of followers was significantly correlated with its number of views, number of people in its circles, and Fortune 500 ranking.[2] That is to say, *the more followers a company had on Google+, the more views it tended to receive, the more people it had in its circles, and the higher it ranked on the Fortune 500 list.*

Companies' Google+ Posts

Using a method similar to what we described in the Twitter, Facebook, and YouTube chapters, we systematically sampled posts from those companies that had Google+ accounts. In particular, we randomly sampled one post for each month for each company, as long as the company had at least one post for that specific month. It is worth pointing out that the time frame for sampling was from November 2011 to December 2013, because Google+ was established in June 2011 and it did not allow companies to create posts until November 2011 (this was different from sampling for the other three platforms, where the time frame was from January 2009 to December 2013). Therefore, a maximum of 26 posts for a company were included using this sampling method.[3]

Once a post was sampled, a screenshot of it was immediately saved in a document. Three metrics associated with each post were recorded and entered in the database, including (1) the number of "plusses" (*reputation*), (2) the number of "comments" (*relationship*), and (3) the number of "shares" (*trust*). These Google+ metrics functioned in a similar way to Facebook's (the number of "likes," the number of "comments," and the number of "shares"). In addition, we recorded the top two comments on each post (if a post had only one comment, we recorded that comment; if it had no comment, we then considered its comment data to be none [0]).

After completing this sampling procedure, we ended up with a total of 2,852 posts. On average, each post generated approximately 32 plusses (minimum: 0; maximum: 2,404), 5 comments (minimum: 0; maximum: 500), and 5 shares (minimum: 0; maximum: 1,089). We conducted correlation analyses to see if these three metrics were significantly associated with each other. The results suggested that they were all significantly and positively correlated.[4] In other words, *when a Google+ post received more plusses, it tended to generate more comments and shares as well.*

Recording the top two comments on each post, we collected a total of 1,505 comments. To transform these comments into numerical measures in the database, we implemented a content analysis procedure similar to what we used for the other three platforms. The two coders from the Twitter project served as the coders for this Google+ project. In the pretest session, we randomly selected five companies that had Google+ accounts and asked the two coders to code the comments on those five companies' posts. Each comment was coded into one of the five categories: (1) compliment, (2) complaint, (3) question, (4) self-promotion, and (5) neutral opinion.[5] A comparison of the two coders' coding showed that the overall agreement was 92.7%, suggesting a satisfactory level of reliability. We then divided the remaining content into two parts, and each coder coded one of them. The whole coding procedure lasted approximately 5 weeks.

The following examples are selected sample comments for each category for illustration purposes. They are all original and we did not correct any spelling or grammar mistakes.

(1) Compliment: "fantastic..........."
(2) Complaint: "Your corporate stores blow. Will never buy a phone there, and little if any help on a Sunday. Extremely upset. Especially, since I pay a lot for my service."
(3) Question: "Where could we find 5 phase stepper motor solution?"
(4) Self-promotion: "follow my page plz ppl"
(5) Neutral opinion: "Health Point"

Similar to what we observed in comments on the other three platforms, users' comments on the Fortune 500 companies' Google+ posts appeared to be mostly positive or neutral (40.7% compliment, 12.2% complaint, 6.2% question, 3.1% self-promotion, and 37.8% neutral opinion). Based on its valence, we assigned a score to each comment (compliment being +1, complaint being –1, and all others being 0). Because we recorded the top two comments on each post, we then averaged the scores for both comments. Regarding each post's overall comment valence (*credibility, confidence*), we used the same formula we used in the other three platforms.

Overall comment valence = number of comments × average comment valence

Companies' Google+ Activities and Their Stock Returns

To test the relationship between a company's Google+ activeness and its business performance from a finance perspective, we conducted a hierarchical regression analysis, with a company's monthly stock return being the consequence variable and its Google+ activeness indicators and the monthly return of the Standard & Poor's 500 Composite Index being the predictor variables. This analysis was similar to what we did with the other three platforms, aiming to test whether a company's Google+ activeness had a significant effect on its monthly stock return, independent of the impact of the Standard & Poor's 500 Composite Index. Specifically, we had six variables in our database for this analysis:

(1) The monthly stock return of each public company,
(2) The monthly return of the Standard & Poor's 500 Composite Index,
(3) The number of plusses (*reputation*) of each Google+ post (it represented a specific month when the post was created on Google+),
(4) The number of comments (*relationship*) on each Google+ post (it represented a specific month when the post was created on Google+),
(5) The number of shares (*trust*) of each Google+ post (it represented a specific month when the post was created on Google+), and
(6) The overall comment valence score (*credibility, confidence*) of each Google+ post (it represented a specific month when the post was created on Google+).

Since several of these variables were significantly correlated with each other, we standardized all of them before entering them in the following two regression equations.[6] By comparing the two equations, we could see whether any of the four Google+ activeness indicators (the numbers of plusses, comments, and shares, and the overall comment valence) had a significant and unique impact on a company's monthly stock return.

Equation 1: $y = a + b_1x_1 + e$, and
Equation 2: $y = a + b_1x_1 + b_2x_2 + b_3x_3 + b_4x_4 + b_5x_5 + e$, where

y = the monthly stock return of each public company
a = constant
x_1 = the monthly return of the Standard & Poor's 500 Composite Index
x_2 = the number of plusses of each Google+ post
x_3 = the number of comments on each Google+ post
x_4 = the number of shares of each Google+ post
x_5 = the overall comment valence score of each Google+ post
b_1 to b_5 = regression coefficients
e = residue

The regression analysis results revealed that a company's monthly stock return was significantly influenced by the Standard & Poor's 500 Composite Index's monthly return.[7] However, none of the regression coefficients associated with a company's Google+ activeness was significant, suggesting that *a company's monthly stock return was not significantly affected by its Google+ nonfinancial activeness.*[8]

Companies' Google+ Activities and Their Profitability

To test the relationship between a company's Google+ activeness and its business performance from an accounting perspective, we conducted six sets of hierarchical regression analyses. The procedure of these analyses was similar to that in the Twitter, Facebook, and YouTube research. Specifically, we used six measures to represent a company's business performance from an accounting perspective, including its quarterly (1) total revenue, (2) net income (or loss), (3) earnings per share, (4) profit margin, (5) return on assets, and (6) return on equity. These measures were either directly recorded from the database Compustat or calculated based on the formula described in earlier chapters, and they served as the consequence variables in the regression equations.

To control for the influence of seasonality and business sector on a company's business performance, we included 12 dummy variables in the regression equations, and they served as predictor variables (the same as those used for the Twitter, Facebook, and YouTube platforms). Finally, we calculated each company's quarterly Google+ activeness by adding its numbers of plusses, comments, and shares, and overall comment valence over 3 months. In this calculation process, we adjusted the formula for those companies that did not end their fiscal year in December. These quarterly Google+ activeness figures also served as predictor variables in the regression equations. To sum up, we entered the consequence variables and predictor variables in the regression equations in the following way (we standardized all the variables except for the dummy variables, because several of them were significantly correlated with each other):[9]

Equation 1: $y = a + b_1x_1 + \ldots + b_3x_3 + b_4x_4 + \ldots + b_{12}x_{12} + e$, and
Equation 2: $y = a + b_1x_1 + \ldots + b_3x_3 + b_4x_4 + \ldots + b_{12}x_{12} + b_{13}x_{13} + b_{14}x_{14} + b_{15}x_{15} + b_{16}x_{16} + e$, where

y = each public company's quarterly total revenue, net income (or loss), earnings per share, profit margin, return on assets, or return on equity
a = constant
x_1 to x_3 = three dummy variables for seasonality
x_4 to x_{12} = nine dummy variables for business sector
x_{13} = the quarterly number of Google+ plusses
x_{14} = the quarterly number of Google+ comments
x_{15} = the quarterly number of Google+ share
x_{16} = the quarterly overall Google+ comment valence score
b_1 to b_{16} = regression coefficients
e = residue

The regression analyses results suggested several significant findings. First, a company's quarterly number of Google+ plusses (*reputation*) had a significant and positive impact on its quarterly earnings per share.[10] Second, a company's quarterly number of Google+ shares (*trust*) had significant and positive effects on its quarterly net income (or loss) and earnings per share.[11] Third, a company's quarterly number of Google+ comments (*relationship*) had significant but *negative* effects on its quarterly net income (or loss), earnings per share, profit margin, and return on assets.[12] Finally, a company's quarterly total revenue and return on equity were not significantly affected by any of its Google+ activeness indicators.[13]

The first two findings were consistent with our expectations. People give a company's Google+ posts "plusses" probably because they favor the company or the information in the posts. Similarly, people share a company's Google+ posts with their friends possibly because they consider the information in the posts to be important to know or interesting to read. Both behaviors (giving plusses to a post and sharing a post) suggest third-party endorsement in terms of trust and reputation, thus helping the company to gain more profits (a higher net income and earnings per share).

However, the third finding was surprising. We suspected that this finding might again be caused by a few outliers in the database. Therefore, we excluded those outliers and reconducted the regression analyses. The negative effects of a company's quarterly number of Google+ comments on its quarterly net income (or loss), profit margin, and return on assets became insignificant with this exclusion of outliers, but *the negative effect on the quarterly earnings per share remained significant*.[14] We further explored why the effect of a company's quarterly number of Google+ comments on its quarterly earnings per share was negative by looking at the correlation between these two variables. It was found that, in fact, they were significantly and positively correlated. Thus, we conclude that the negative effect detected in the regression analyses was caused by a collinearity problem, and we suggest that readers interpret this finding with caution.[15]

Moreover, we examined whether a company's Google+ nonfinancial activeness would differ when it was profitable as opposed to when it was not. In particular, we performed a series of independent t-tests to compare two types of quarters (profitable vs. nonprofitable) on the numbers of Google+ plusses, comments, and shares, and the average comment valence. The criterion used to determine whether a specific quarter was profitable or not was the same for the Twitter, Facebook, and YouTube platforms (based on a company's reported quarterly net income or loss). No significant difference was detected.[16] In other words, *a company's Google+ activeness in a profitable quarter did not significantly differ from that in a nonprofitable quarter*, measured in its quarterly numbers of Google+ plusses, comments, shares, and the average comment valence.

Companies' Google+ Activities Across Business Sectors

Finally, we explored whether different business sectors significantly differed with regard to their Google+ activeness. The 10 business sectors that we included in these analyses were the same we included in the research on the Twitter, Facebook, and YouTube platforms. Specifically, we conducted four one-way ANOVA tests to examine the differences among the 10 business sectors regarding their quarterly numbers of Google+ plusses, comments, and shares, and the average comment valence.

According the ANOVA test results, there were not many differences among various business sectors regarding their quarterly numbers of Google+ plusses,[17] comments,[18] and shares.[19] The *information technology sector was more successful in generating plusses, comments, and shares* (plusses: approximately 213; comments: approximately 33; shares: approximately 46), *especially when compared to the industrials and financials sectors* (plusses: ranged from approximately 10 to 46; comments: ranged from approximately 2 to 8; shares: ranged from approximately 2 to 6).

In regard to the average comment valence, it was found that the comments on the posts by the *consumer discretionary, industrials, and consumer staples sectors were significantly more positive* (average comment valence scores ranged from approximately 0.45 to 0.54) *than those by the financials, health care, and telecommunication services sectors* (average comment valence scores ranged from approximately –0.20 to –0.01).[20]

Limitations of Our Data

Google launched Google+ in 2011 to compete with Facebook.[21] Although its history is much shorter compared to other established social media platforms such as Facebook and Twitter, Google+ gained popularity fairly quickly. We intentionally included this "latecomer" in the book because we believe that an examination of how the Fortune 500 companies use Google+ could bring new and unique perspectives. However, we also realize that we could not collect as much data from Google+ as we could from Twitter, Facebook, and YouTube. As noted earlier in this chapter, we could sample a maximum of just 26 posts from a company that had a Google+ account (as opposed to a maximum of 60 tweets, posts, or videos from a company that had a Twitter, Facebook, or YouTube account). This shortcoming raised an issue in our data analyses in this chapter regarding statistical power. Because the sample size for the Google+ platform was smaller than those for the Twitter, Facebook, and YouTube platforms, we might not be able to detect an effect in the data analyses (which might be detected with a larger sample size).[22]

Synopsis of Research Findings

In this Google+ research project, we adopted data collection and data analysis methods similar to those we used for the other three platforms (Twitter, Facebook, and YouTube). Each company's nonfinancial activities on Google+ were investigated. The relationship between its Google+ activeness and its business performance was also tested. The key findings are synopsized as follows:

- Google+ appeared to be a valuable platform for corporate communication, since 367 companies on the Fortune 500 list had established Google+ accounts, though this adoption rate was somewhat lower than those for Twitter, Facebook, and YouTube.

- A company's total number of Google+ followers was significantly correlated with its number of views, the number of people in its circles, and its Fortune 500 ranking. *The more followers a company had, the more views it tended to generate, the more people it was likely to have in its circles, and the higher it ranked on the Fortune 500 list.*
- The public showed a reasonable level of engagement with the Fortune 500 companies on Google+ by reading the companies' posts, giving plusses to the posts, commenting on the posts, and sharing the posts. On average, each Google+ post generated about 32 plusses, 5 comments, and 5 shares.
- A Google+ post's numbers of plusses, comments, and shares were significantly correlated. In other words, *when a post gets "popular," it tends to generate more plusses, comments, and shares.*
- Similar to what we observed for the Twitter, Facebook, and YouTube platforms, we found that when the public commented on posts by the Fortune 500 companies on Google+, most of the time they were either positive or neutral.
- A company's Google+ nonfinancial activeness (including the numbers of plusses, comments, and shares, and the overall comment valence) did *not* have a significant effect on its monthly stock return.
- A company's quarterly number of Google+ plusses (*reputation*) was found to have a significant and positive impact on its quarterly earnings per share.
- A company's quarterly number of Google+ shares (*trust*) had significant and positive effects on its quarterly net income (or loss) and earnings per share.
- We found that a company's quarterly number of Google+ comments (*relationship*) had significant but *negative* effects on its quarterly net income (or loss), earnings per share, profit margin, and return on assets. We suggest that readers interpret these effects with caution, because they were likely to be caused by a few outliers in the database and a collinearity issue in the regression analyses.

- A company's Google+ nonfinancial activeness in a profitable quarter *did not* significantly differ from that in a nonprofitable quarter.
- There were certain differences among various business sectors in regard to their Google+ activeness. *The information technology sector seemed to be more successful in generating plusses, comments, and shares, compared to the industrials and financials sectors.* Furthermore, *the public's comments on the posts by the consumer discretionary, industrials, and consumer staples sectors were found to be significantly more positive than those on the posts by the financials, health care, and telecommunication services sectors.*

Summary

In this chapter we examined how the Fortune 500 companies used Google+ for their corporate communication from late 2011 to the end of 2013. Again, Google+ was found to be a useful social media platform for companies to engage their stakeholders online. A company's Google+ activeness had a somewhat significant relationship with its business performance. In the next chapter, we will combine the data collected from all four social media platforms (Twitter, Facebook, YouTube, and Google+) and analyze and discuss them from a more theoretical standpoint.

Chapter Eight

The Big Four

Chapter Overview

In the previous chapters we have presented our research findings on each of the four social media platforms. In this chapter, we will combine all the data that we collected and analyze them at a more aggregate and more theoretical level. Since the data cover a 5-year span from January 2009 to December 2013, we will also test the social media usage trends across different years. In conducting such data analyses, we hope to answer the following research questions:

- What is the best way to measure a company's outcomes on social media, including reputation, relationship, trust, credibility, and confidence?
- Do a company's social media outcomes affect its business performance from a finance perspective, measured in its monthly stock return?
- Do a company's social media outcomes affect its business performance from an accounting perspective, measured in its quarterly

total revenue, net income (or loss), earnings per share, profit margin, return on assets, and return on equity?

- Do a company's social media outcomes differ across various business sectors?
- Does a company's social media activeness differ across the four platforms Twitter, Facebook, YouTube, and Google+?
- Does a company's overall presence on social media (regardless of activeness) differ across various business sectors?
- Are there significant differences across 5 years (from 2009 to 2013) with regard to how a company adopts social media for corporate communication and how the public responds to its communication messages on social media?

Social Media Nonfinancial Outcomes

As described in Chapters 1 and 2, a company's nonfinancial outcomes include five communication variables related to public relations and corporate communication: reputation, relationship, trust, credibility, and confidence. These are five broad concepts that need to be further "operationalized" so that their effects can be tested.[1] Specifically, we need to give a well-tailored definition for each concept and create a formula based on that definition for data calculation. Each formula should precisely reflect the conceptual meaning of each outcome. Because this book is focused on the Fortune 500 companies' usage of social media for corporate communication, we thus define each nonfinancial outcome in a social media context. First, we define a company's reputation on social media in terms of the level of favorable opinions the public expresses in response to the company's communication messages, reflected in the numbers of Twitter favorites, Facebook likes, YouTube likes, and Google+ plusses. Accordingly, we created a new variable in our database, "*reputation*," using the following formula for calculation:

Reputation = (number of Twitter favorites + number of Facebook likes + number of YouTube likes + number of Google+ plusses)/4

Second, we define a company's relationship in terms of the amount of interaction the public has with the company on social media, measured in the numbers of Twitter replies, Facebook comments, YouTube comments, and Google+ comments. We thus created a new variable, *"relationship,"* using the following formula:

Relationship = (number of Twitter replies + number of Facebook comments + number of YouTube comments + number of Google+ comments)/4

Third, we define a company's trust in terms of the public's willingness to share the company's communication messages with others. Such willingness can be reflected in the public's sharing activities on social media. We used the following formula to create a new variable, *"trust,"* in our database:

Trust = (number of Twitter retweets + number of Facebook shares + number of YouTube shares + number of Google+ shares)/4

Next, we believe that a company's credibility can be demonstrated by how active the public is in seeking a connection with the company on social media. Such connection activities include following a company on Twitter, subscribing to a company's YouTube channel, and following a company on Google+. We thus created the following formula and used it to calculate a new variable, *"credibility"*:

Credibility = (number of Twitter followers + number of YouTube subscribers + number of Google+ followers)/3

Finally, we think that the confidence a company gains from the public is reflected in the valence of the public's interactions with the company. The more positive feedback the public gives to the company, the more confidence they demonstrate. Therefore, we used the following formula to measure a company's *"confidence"*:

Confidence = (overall Twitter reply valence + overall Facebook comment valence + overall YouTube comment valence + overall Google+ comment valence)/4

After we created these five new variables, we conducted correlation analyses among them. It was found that they were all significantly and positively correlated, except for the relationship between trust and confidence.[2] In other words, *when a company establishes a good "relationship" with its public, the public tends to "trust" the company and confer on it a high "reputation" and "credibility." Consequently, the public gains more "confidence" in the company.*

Moreover, we performed a correlation analysis between a company's Fortune 500 ranking and its social media outcomes. Significant correlations were discovered among a company's Fortune 500 ranking and its reputation, trust, and credibility.[3] That is to say, the *companies ranked higher on the Fortune 500 list were likely to be considered as more reputable, trustworthy, and credible.*

Social Media Nonfinancial Outcomes and Business Performance

To test whether a company's relationship, reputation, trust, credibility, and confidence affect its business performance, we used the company's financial data and accounting data as judgmental criteria, similar to what we did in the previous four chapters. First, we tested whether the five outcomes (relationship, reputation, trust, credibility, and confidence) affected a company's monthly stock return. Specifically, we used a hierarchical regression method to examine whether any of the outcome measures had an effect on a company's monthly stock return, independent of the impact of the Standard & Poor's 500 Composite Index's monthly return. The regression equations for these analyses are listed as follows (all variables in the regression equations were standardized):

Equation 1: $y = a + b_1x_1 + e$, and
Equation 2: $y = a + b_1x_1 + b_2x_2 + e$, where

y = a company's monthly stock return
a = constant

x_1 = the Standard & Poor's 500 Composite Index's monthly return
x_2 = a company's monthly reputation, relationship, trust, credibility, or confidence
b_1 and b_2 = regression coefficients
e = residue

Based on the five sets of hierarchical regression analyses, we found that a company's monthly stock was significantly affected by the monthly return of the Standard & Poor's 500 Composite Index.[4] However, *none of the outcome measures showed a significant effect on a company's monthly stock return.*[5]

Next, we examined the effects of relationship, reputation, trust, credibility, and confidence on a company's business performance from an accounting perspective. In particular, we conducted a series of hierarchical regression analyses to see whether a company's relationship, reputation, trust, credibility, and confidence would influence its quarterly total revenue, net income (or loss), earnings per share, profit margin, return on assets, and return on equity, independent of the effect of seasonality and business sector. The specific regression equations that were tested are listed as follows (all variables in the regression equations were standardized except for those dummy variables):

Equation 1: $y = a + b_1x_1 + \ldots + b_3x_3 + b_4x_4 + \ldots + b_{12}x_{12} + e$, and
Equation 2: $y = a + b_1x_1 + \ldots + b_3x_3 + b_4x_4 + \ldots + b_{12}x_{12} + b_{13}x_{13} + e$, where

y = a company's quarterly total revenue, net income (or loss), earnings per share, profit margin, return on assets, or return on equity
a = constant
x_1 to x_3 = three dummy variables for seasonality
x_4 to x_{12} = nine dummy variables for business sector
x_{13} = a company's quarterly reputation, relationship, trust, credibility, or confidence
b_1 to b_{13} = regression coefficients
e = residue

Several significant findings were discovered. First, a company's reputation had a significant and positive effect on both the net income (or loss) and profit margin.[6] Second, a company's relationship had a significant and positive effect on its earnings per share.[7] Third, trust affected the net income (or loss), earnings per share, and profit margin significantly and positively.[8] Finally, credibility was found to influence the total revenue, net income (or loss), profit margin, and return assets in a significant and positive way.[9] No significant relationship was found between confidence and the six business performance measures.[10] Based on these results, we concluded that *a company's nonfinancial outcomes were indeed significantly associated with its business bottom-line outcomes such as the total revenue, net income (or loss), earnings per share, profit margin, and return on assets.*

Social Media Nonfinancial Outcomes Across Business Sectors

Another objective of our data analyses in this chapter was to examine whether different business sectors differed in their social media outcomes. We conducted five sets of one-way ANOVA tests for this purpose. The 10 business sectors that we included in these analyses were the same as those analyzed in the previous four chapters. The business sectors' quarterly scores on reputation, relationship, trust, credibility, and confidence were standardized and then compared. *Significant differences were detected for reputation, relationship, trust, and credibility, while no significant difference was found for confidence.*[11] As seen in Figure 8.1, the consumer discretionary sector appeared to be a clear "winner" in using social media for corporate communication, based on the test results. Companies in this sector apparently had more two-way interactions with their publics on social media, and their communication messages were significantly shared more and liked more. With regard to the overall credibility, information technology, consumer discretionary, and consumer staples were significantly stronger than other business sectors because they tended to attract more followers or subscribers on social media.

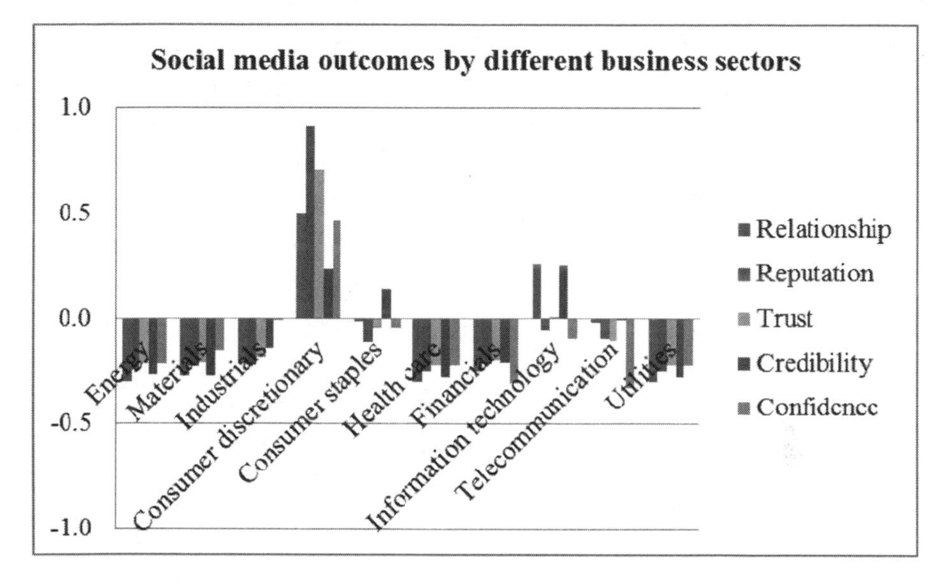

Figure 8.1. Differences in social media outcomes across various business sectors.

Social Media Activeness Across Platforms

We also intended to examine whether a company's social media activeness differed across various platforms. In other words, did the Fortune 500 companies consider certain social media platforms to be more (or less) important than others for their corporate communication, as reflected in their activeness on those platforms? To address this research question, we conducted a series of correlation analyses.

First, we conducted a correlation analysis with a company's monthly Twitter replies, Facebook comments, YouTube comments, and Google+ comments.[12] It was found that they were all significantly and positively correlated. *The strongest correction was found between Facebook comments and Google+ comments* (correlation coefficient $r = 0.52$), and *the weakest correlation was between Twitter replies and YouTube comments* (correlation coefficient $r = 0.05$).

Second, we performed a correlation analysis with a company's monthly Twitter favorites, Facebook likes, YouTube likes, and Google+ plusses.[13] All the correlations were found to be significant and positive, except for that between the numbers of Facebook likes and YouTube likes. *The strongest correlation was between Twitter favorites and Google+*

plusses (correlation coefficient *r* = 0.57), whereas *the weakest correlation was between Twitter favorites and YouTube likes* (correlation coefficient *r* = 0.06).

Third, we tested the correlations among a company's monthly Twitter retweets, Facebook shares, YouTube shares, and Google+ shares.[14] The results showed that *most of the correlations were significant and positive, and their strengths were similar* (correlation coefficient *r* ranged from 0.17 to 0.30). However, the number of YouTube shares was not significantly correlated with the number of Twitter retweets and the number of Facebook shares.

Next, we examined the correlations among the numbers of Twitter followers, YouTube subscribers, and Google+ followers.[15] We found that they were *all significantly and positively correlated, and the strength of those correlations appeared to be moderate* (correlation coefficient *r* ranged from 0.27 to 0.46).

Finally, we conducted a correlation analysis with a company's monthly average valence scores of Twitter replies, Facebook comments, YouTube comments, and Google+ comments.[16] The *Facebook comments' average valence was significantly and positively correlated with those of Twitter replies and YouTube comments* (both correlations were somewhat weak, with correlation coefficient *r* being 0.10 and 0.07, respectively), but no other significant correlation existed.

Based on these findings, we concluded that the Fortune 500 companies actively used all four social media platforms—Twitter, Facebook, YouTube, and Google+. Their nonfinancial activeness on one social media platform was in fact significantly correlated with activeness on the other three platforms. Thus, *all four platforms were deemed as important for corporate communication.*

Social Media Presence Across Business Sectors

Furthermore, we examined whether various business sectors differed in their adoption rates of social media. Specifically, we classified the adoption rates of social media into five levels: (1) no presence on any of the four social media platforms (Twitter, Facebook, YouTube, and Google+), (2) presence on one of the four platforms, (3) presence on two of the four platforms, (4) presence on three of the four platforms, and (5) presence on all four platforms. We then performed a chi-square test.[17] The purpose of the test was to reveal whether there existed significant differences among various business sectors with regard to their specific levels of social media adoption. As seen in Figure 8.2, significant differences were detected.[18] In particular, information technology was the business sector with the highest level of social media adoption (93.8% of information technology companies had a presence on all four social media platforms). The telecommunication services sector was another where social media usage was quite popular (80.0% of telecommunication services companies had a presence on all four social media platforms, and 20.0% had a presence on three platforms). In contrast, energy was the business sector where social media were used the least (28.2% of energy companies had no social media presence, and only 23.1% had a presence on all four social media platforms). The other business sectors were somewhere between these two extremes.

Moreover, we conducted a one-way ANOVA test to see whether there was a significant association between a company's social media adoption rate and its Fortune 500 ranking. The test results suggested no significant connection.[19] That is to say, the *companies that ranked higher on the Fortune 500 list did not significantly differ from those that ranked lower, in regard to their presence on social media.*

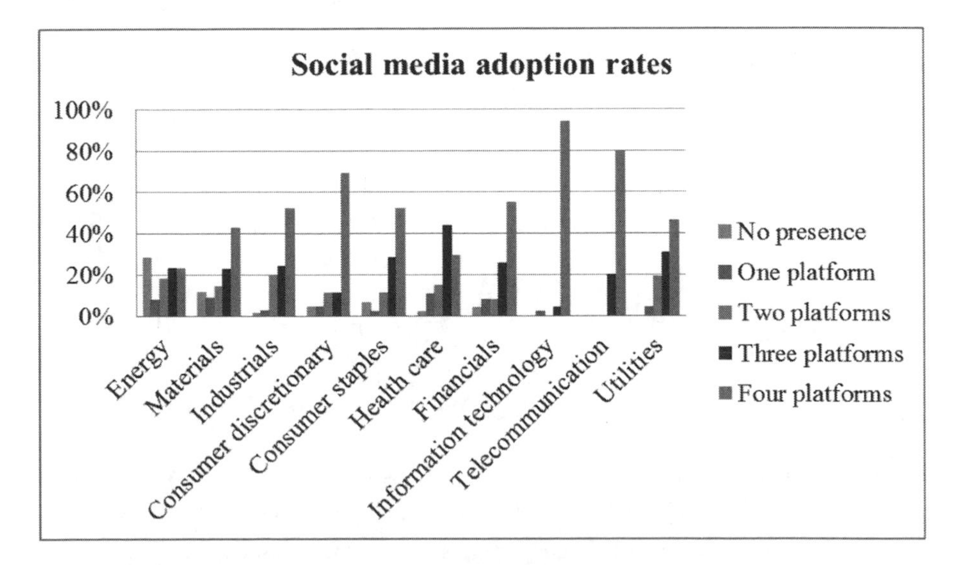

Figure 8.2. Social media adoption rates across various business sectors.

Yearly Trends

Finally, we analyzed our data to see whether there were significant trends across different years. We created a variable called "time" in the database to reflect the month when a company posted a specific message on a certain social media platform (Twitter, Facebook, YouTube, or Google+). Since the time frame of our research was from January 2009 to December 2013 (a total of 60 months), this time variable had 60 levels of value from 1 to 60, with 1 representing December 2013 and 60 representing January 2009.

To test the trend of how the Fortune 500 companies used Twitter over the 5 years, we conducted a correlation analysis with a company's monthly Twitter replies, Twitter favorites, and Twitter retweets, and the time variable. It was found that Twitter replies, Twitter favorites, and Twitter retweets were all significantly correlated with time.[20] As seen in Figure 8.3, as time went by, the public tended to be more enthusiastic about a company's nonfinancial messages on Twitter, by more often replying to them, marking them as favorites, and retweeting them.

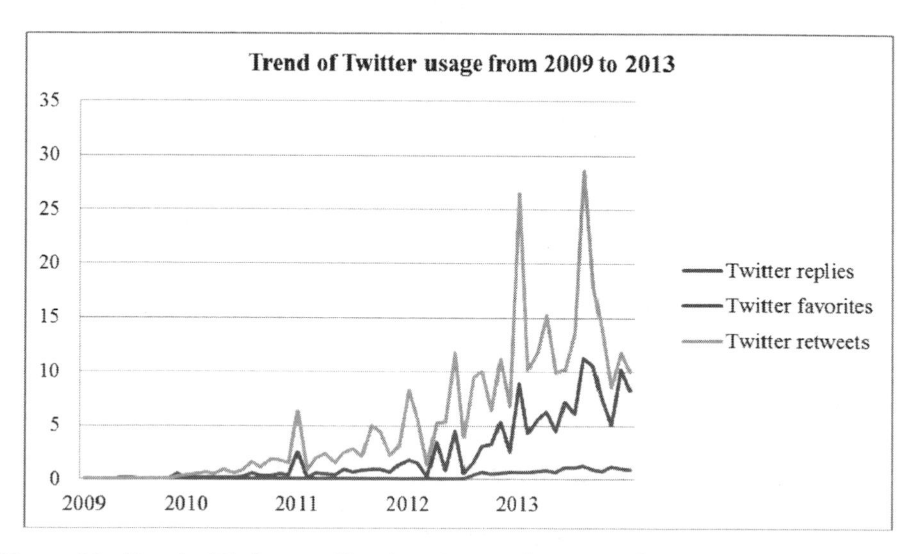

Figure 8.3. Trend of Twitter replies, favorites, and retweets from 2009 to 2013.

Next, we performed a correlation analysis with a company's monthly Facebook comments, likes, and shares, and the time variable to examine the trend of Facebook usage. Again, significant correlations were detected.[21] As shown in Figure 8.4, over the years, the public became more involved in interacting with the companies on Facebook, by more often commenting on, liking, and sharing their Facebook posts.

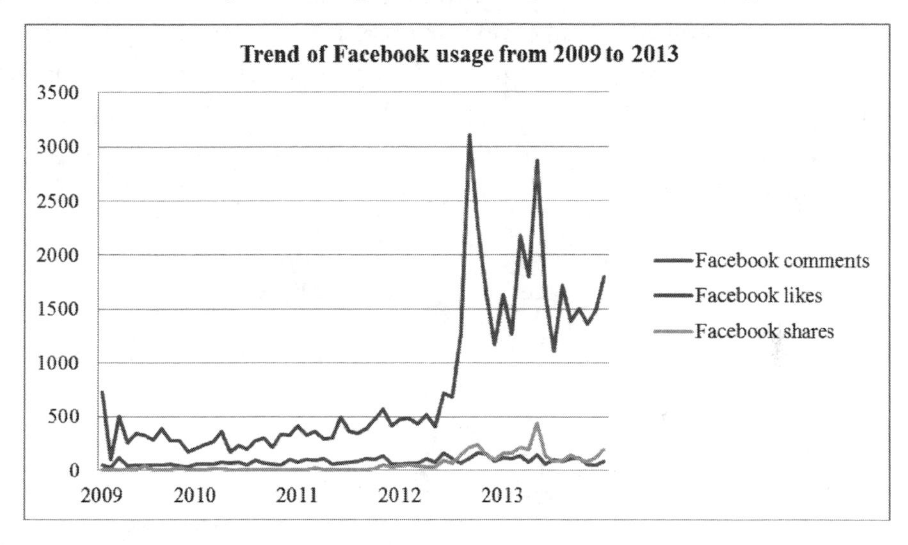

Figure 8.4. Trend of Facebook comments, likes, and shares from 2009 to 2013.

Furthermore, we looked at the correlations between a company's monthly YouTube video views, comments, likes, dislikes, and shares, video lengths, and the time variable. Three significant correlations were found.[22] As seen in Figure 8.5, as time progressed, the videos that the Fortune 500 companies put on YouTube tended to be longer, and the public more frequently liked and disliked those videos.

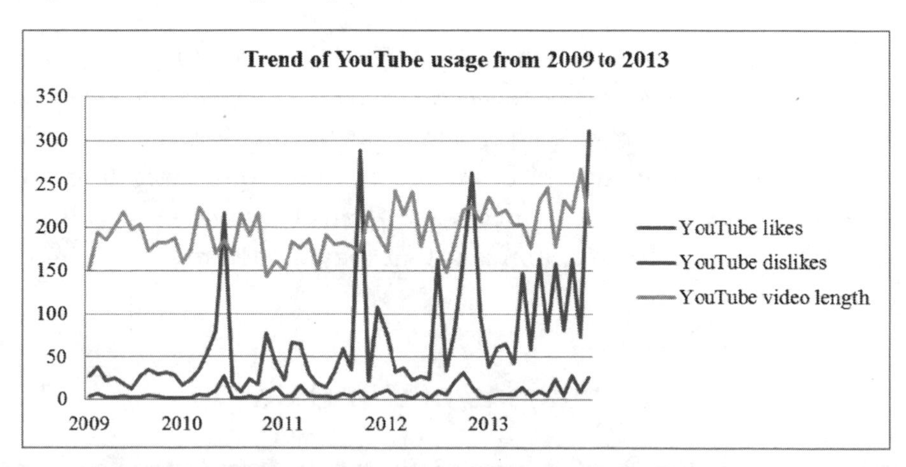

Figure 8.5. Trend of YouTube likes, dislikes, and video lengths (in seconds) from 2009 to 2013.

Finally, we examined the trend of Google+ usage by conducting a correlation analysis with a company's monthly Google+ comments, plusses, and shares, and the time variable. It was found that Google+ comments and plusses were significantly correlated with the time, but their correlation directions were different.[23] As illustrated in Figure 8.6, from late 2011 to the end of 2013, the public tended to give more plusses to but fewer comments on a company's Google+ posts.

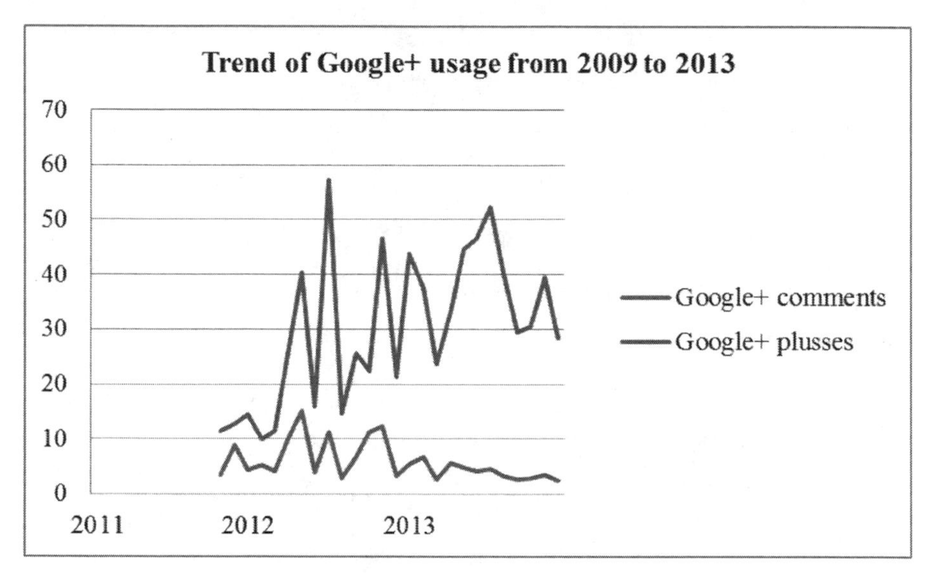

Figure 8.6. Trend of Google+ comments and plusses from 2011 to 2013.

Taken together, we concluded that *the general trend from 2009 to 2013 was that the public becomes more engaged with interacting with companies on social media*. As time progressed and Internet technologies became more mature, the public tended to be more active in communicating with companies on various social media platforms by commenting, liking, and sharing (the only exception was that they tended to provide fewer comments on Google+).

Limitations of Our Data

As presented earlier, significant associations were found between a company's bottom-line business performance measures and its reputation, relationship, trust, and credibility on social media. However, no significant relationship was discovered between the public's *confidence* in a company and its business performance. We suspect that this insignificant result was caused by the limitations of our data. Specifically, the confidence variable in our database was calculated as the average valence of Twitter replies, Facebook comments, YouTube comments, and Google+ comments. As noted in the previous four chapters, we recorded the top two replies/comments on a message and content analyzed their valence. This method could provide a reasonable snapshot of how the public responded to a company's communication messages on social media, but the valence score generated might not be 100% accurate. First, the amount of replies/comments on each social media platform differed to a certain extent (Twitter: an average of 0.43 replies; Facebook: an average of 87.92 comments; YouTube: an average of 12.64 comments; Google+: an average of 5.39 comments). Therefore, when we calculated the valence scores of the top two replies/comments on a specific message, there was a chance that it might be an overestimation or an underestimation. Second, although we implemented a consistent content analysis procedure in all cases to ensure that our coders' coding was reliable, the valence scores generated from this method were categorical. As discussed in Chapter 2, this type of categorical data, by nature, might not be as accurate as continuous data.

Given that our data had these limitations, we suggest that readers keep these in mind when interpreting the findings associated with the nonfinancial confidence variable. Insignificant results did not necessarily mean that the public's confidence in a company had no impact on its business performance. On the contrary, based on the theoretical framework, we believe that confidence is an important factor that will affect how a company performs in the business world. We hope that in the future more scholars and professionals will join us in this research direction and provide more empirical tests on such a proposition.

Synopsis of Research Findings

We analyzed the data from a more theoretical perspective in this chapter by looking at the associations between a company's nonfinancial outcomes on social media and its bottom-line business performance measures. We also examined whether various business sectors differed in their social media outcomes and their overall social media adoption rates, as well as their activeness across four social media platforms. Finally, we explored the social media usage trends from 2009 to 2013. Many significant findings were detected based on our data analysis results. The key points are as follows:

- A company's *reputation* on social media had significant and positive effects on its net income (or loss) and profit margin.
- The *relationship* a company established on social media had a significant and positive effect on its earnings per share.
- The *trust* a company gained on social media had significant and positive effects on its net income (or loss), earnings per share, and profit margin.
- A company's *credibility* on social media had significant and positive effects on its total revenue, net income (or loss), profit margin, and return on assets.
- The public's *confidence* in a company on social media had no significant effect on its bottom-line business performance.
- A company's *nonfinancial outcomes* on social media including reputation, relationship, trust, credibility, and confidence had no significant effect on its monthly stock return.
- There existed significant differences among 10 business sectors regarding their social media outcomes. The consumer discretionary sector was the most successful in generating *reputation, relationship*, and *trust* on social media. Three sectors—information technology, consumer discretionary, and consumer staples—were deemed as more *credible* on social media than other sectors.
- A company's *nonfinancial activeness* on the four social media platforms Twitter, Facebook, YouTube, and Google+ was significantly and positively correlated. *When a company was active on one platform, it tended to be active on the other three platforms as well.*
- The information technology and telecommunication services were the two business sectors for which adopting social media

for corporate communication was very popular. In contrast, the social media adoption rate was the lowest in the energy sector.

- From 2009 to 2013, as Internet technologies became more mature, the public tended to be more enthusiastic and active in communicating with companies on social media. They tended to like (or dislike) and share a company's social media messages more frequently. They were also more likely to provide comments (except for on Google+).

Summary

This chapter builds upon the research detailed in the previous four chapters, presenting a more theory-driven data analysis approach. The social media metrics that we sampled from different platforms were included in the same theoretical framework for analysis and interpretation. Based on the research findings, the simple conclusion is that a company's social media activeness is significantly and positively associated with its business bottom-line measures. In the next chapter, we will provide a more detailed interpretation and analysis of all the research findings presented in Chapters 4 to 8.

Chapter Nine

Research Implications

Chapter Overview

In this chapter we will revisit and summarize all the research findings presented in the previous five chapters. We aim to interpret those research findings from both theoretical and practical standpoints, and provide implications for businesses. Specifically, we want to address the following questions:

- What are the three broad research questions that we intend to answer by conducting this research?
- What is the theoretical background of this research?
- How should the research findings be interpreted from a theoretical viewpoint?
- How should the research findings be interpreted from a practical viewpoint?
- What are the major limitations associated with this research?

A Revisit of Our Research Questions

We began this project with several questions in mind. First, we wondered if top companies actually use social media. Second, in exploring company activities on four social media platforms we felt best represented historical social media practices, we examined whether they had any impacts on the financial outcomes of top companies. Third, we wondered if a model of public relations/corporate communication developed in the late 1990s and redefined in 2010–2011 might explain how companies' communication with stakeholders (broadly defined as anyone using any of the four social media platforms) impacted on company financial and accounting outcomes. Perhaps by now the questions we asked seem naïve, but no one had addressed them in such detail before. As such, we would define this project as exploring the basic questions underlying corporate social media use. Before going into specific results and their implications, we need to answer these three questions. *The basic answer to all three is a resounding* YES, *but some of the findings are counterintuitive*. This chapter seeks to frame the project with a theoretical focus, based on the early works on public relations/corporate communication discussed in Chapters 1 and 2.

The Role of Social Media in Corporate Communication

Earlier in this book we argued that social media's role in public relations and corporate communication in particular has created new opportunities to communicate with stakeholders, including employees, customers, stockholders, regulators, and so forth. The major change came in the way communication moved from a one-way model of communication, where a great amount of that communication focused on advertising, to a two-way model of communication, whereby the stakeholder acquired through social media a means of impacting on the enterprise's financial outcomes as a partner in the communication. Indeed, what Grunig and associates (Grunig

& Grunig, 1992; Grunig, Grunig, & Dozier, 2002) called "two-way symmetrical communication," or party-to-party communication with each having an impact on the other, became the norm in the early 2010s. These were later redefined as "nonfinancial" indicators (Weiner, 2006).

This led Stacks (2002) to suggest that from a communication point of view, independent sets of variables were primarily driving enterprise outcomes, particularly return on investment (ROI). As Figure 9.1 indicates, the potential predictors are *mediated* by stakeholder perceptions and expectations (ROE), which in turn influence return on investment (ROI). ROI, then, can take two business-related financial perspectives: *stock prices* as a global indicator of business success or failure, and *accounting practices* as an indicator of impact on the economic impact of marketing strategies on ROE.

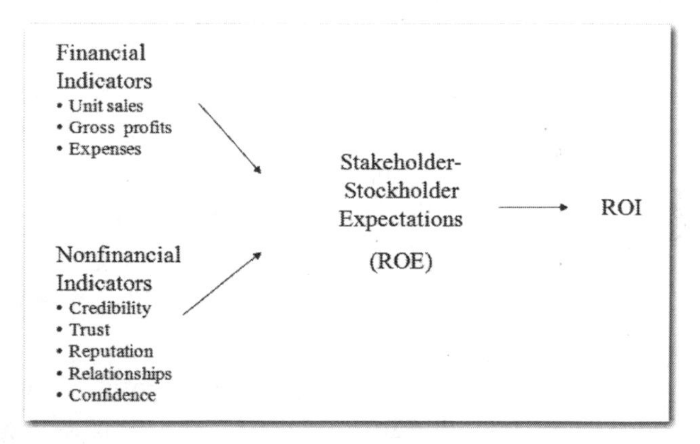

Figure 9.1. The initial model of ROE and ROI.

In 2010 Stacks modified his model of communication influence on ROI. In his 2002 *Primer of Public Relations Research,* he pointed to the communication variables that he argued were "owned" by the communication function in an enterprise: *credibility, relationship, reputation,* and *trust*. Based on the economic downturn of the mid-2000s, he added a fifth variable, *confidence*, as a modifier of communication outcome (see Figure 9.2). Note that each of the initial four variables is interrelated; each influences the other, but each can be manipulated with particular outcomes in mind.

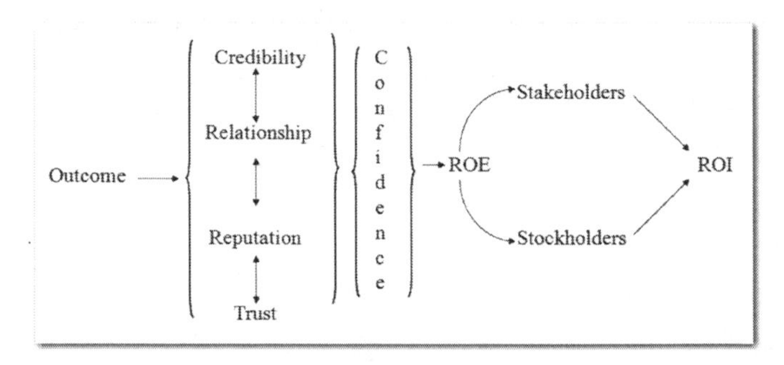

Figure 9.2. The revised model of ROE and ROI.

As noted earlier, the advent of the Internet and the social media platforms that have evolved over the last 25 years provide a way for businesses to communicate directly with stakeholders (initially customers, but since widened to include any stakeholder who might influence overall business outcomes) and vice versa. Hence, a new day in communication arrived whereby stakeholder and enterprise became intertwined in that two-way symmetrical communication model discussed above. In this model, communication (or lack of communication) could be used to impact on one, the other, or both. Further, this new communication model did not allow for message control by the company—stakeholder messaging now could enhance or detract from a company's credibility, relationships, reputation, and trust from the stakeholder's perspective.[1]

What evolved was the impact of a communication agent from the company's perspective who could make the company's arguments for it. This, as noted in the opening chapters, reflects the concept of a "third-party endorser" who is supposedly at arm's length from the company. On the stakeholder side, other stakeholders become endorsers for the company. However—and this is a big however—stakeholder commentary through social media platforms cannot be controlled, and is seen as "earned" by other stakeholders, as the communication often relates to individual experiences with products, services, and company representatives.[2] Thus, such "testimony" is often seen as more credible and trustworthy than the company's communication, and as having an impact on others' perceptions of that company's credibility and reputation.[3] The outcome normally will be a "hit" on the company's financial outcomes.

The variables of importance to company financial success are, of course, stock price for publicly traded companies and general accounting procedures and outcomes for all companies. In this book we examined *stock price* for the Fortune 500 companies as one outcome variable, arguing that being on this list demonstrated that the companies were successful, and accounting processes that are often used to understand and predict stock price or value (monthly stock return, quarterly total revenue, quarterly net income or loss, quarterly earnings per share, quarterly profit margin, quarterly return on assets, and quarterly return on equity). Data on these outcome variables were collected over a 5-year period to rule out any strange fluctuations in the data due to environmental or other impacts. In the following section, we revisit the research findings presented in Chapters 4 to 8 and further explain what industrial implications can be drawn based on those findings.

Corporate Use of Social Media

First of all, we found that the companies on the 2013 Fortune 500 list were high users of social media. Twitter and Facebook were the social media platforms used most (84% and 81%, respectively). YouTube and Google+ were used less (78% and 73%, respectively), but still, use of both platforms was high. The lowest use percentage, for Google+, may be due in part to the platform's relative newness. Twitter and Facebook use allow companies and stakeholders to engage in strings of posts that operate much more like normal daily conversation than YouTube allows; again, Google+ bears watching in the future as its use increases or decreases.

Although a large number of companies were active on the four social media platforms, their intensity of use varied among platforms. With Twitter, for instance, some companies did not tweet at all, while others tweeted quite frequently. Facebook use also varied in intensity, with some companies engaging stakeholders from as early as 2007 and others just starting in 2014. There were certain differences among the Fortune 500 companies regarding their YouTube usage history. Some companies started to deliver videos to their publics via YouTube as early as 2005, but some others did not start doing so until 2013. With regard to using Google+, a company's total number of Google+ followers was found to be significantly correlated with its number of views and the number of people in its circles.

In terms of engagement, each platform engaged its users; Facebook had the most engagement. On average, a company's Facebook post generated about 960 likes, 88 comments, and 81 shares. A company's Twitter tweet generated an average of about 4 favorites, 8 retweets, and less than 1 comment. For YouTube and Google+, the level of public engagement was also reasonably high. On average, each Fortune 500 company that had a YouTube channel generated 385 videos, 14,238 subscribers, and 9,314,198 video views. The YouTube videos they created were an average of 3.5 minutes long, and each one averaged approximately 33,989 views, 87 likes, 9 dislikes, 13 comments, and 40 shares. Finally, a company's total number of Google+ followers was significantly correlated with its number of views, the number of people in its circles, and its Fortune 500 ranking. The more followers a company had, the more views it tended to generate, the more people it was likely to have in the circles, and the higher it ranked on the Fortune 500 list; each Google+ post averaged 32 plusses, 5 comments, and 5 shares.

As might be expected, a company's total number of tweets as a measure of relationship was positively associated with its total numbers of following and followers on Twitter. The numbers of likes, comments, and shares generated by a Facebook post were positively correlated, which means that when a company's post got attention from the public, they tended to like it, comment on it, and also share it. A company's YouTube usage history and its number of subscribers were significantly correlated with its Fortune 500 ranking; the higher it ranked, the earlier it tended to have begun to use YouTube and the more subscribers it was likely to have. Finally, a Google+ post's numbers of plusses, comments, and shares were significantly correlated. In other words, when a post got popular, it tended to generate more plusses, comments, and shares.

In conclusion, all four platforms demonstrated relatively high levels of relationship engagement, measured by tweets, retweets, favorites, replies, friends, comments, likes, and so forth. Replies and comments were generally neutral to positive. However, YouTube garnered negative feelings when the comment function was disabled.

Social Media Use Across Business Sector and Year

We also wondered if business sectors would differ in terms of their social media use. For Twitter, telecommunication service companies generated significantly more replies than other types of businesses did, consumer discretionary companies were most successful in generating retweets and favorites, and finally, the stakeholder replies to the tweets of consumer staples and consumer discretionary companies were significantly more positive than their replies to tweets of utilities, energy, and telecommunication service companies. Also, different business sectors differed with regard to their Facebook activeness. The consumer discretionary sector generated significantly more likes on Facebook, while the information technology sector was the most successful in generating comments. Regarding the number of shares, Facebook posts from the industrial sector were shared significantly fewer times than those from the information technology and consumer discretionary sectors; finally, the telecommunications services sector was significantly lower than the other sectors and significantly more negative in the average comment valence to its posts.

YouTube videos produced by the information technology sector were significantly longer than the videos produced by other sectors. The information technology sector also appeared to be the most successful in generating video views, likes, dislikes, and comments. Regarding video comment valence, the viewers' comments on the videos produced by the consumer discretionary sector were the most positive. Furthermore, there were certain differences among various business sectors in regard to their Google+ activeness. The information technology sector was more successful in generating plusses, comments, and shares, compared to the industrials and financials sectors. The public's comments on the posts by the consumer discretionary, industrials, and consumer staples sectors were found to be significantly more positive than those on the posts by the financials, health care, and telecommunication services sectors.

Finally, information technology and telecommunication services were the two business sectors for which adopting social media for corporate communication was very popular. In contrast, the social media adoption rate was the lowest in the energy sector. From 2009 to 2013, as Internet technologies became more mature, the public tended to be more enthusiastic and active in communicating with companies on social media. They tended to like (or dislike) and share a company's social media messages more frequently. They were also more likely to provide comments (except on Google+).

Corporate Social Media Use and Business Performance

When we tested the relationship between a company's social media activeness and its monthly stock return, we found a significant association only between stock price and Facebook. In this instance, the average number of comments a company generated on Facebook each month had a significant and positive impact on the company's monthly stock return.

As we noted earlier, a company's stock price is but one financial variable that can be impacted by social media activity. Analyses of general accounting processes obtained some significant relationships for each social media platform. For Twitter, the number of retweets was significantly and *positively* associated with the net income (or loss), profit margin, and return on assets; when a company had more retweets, it tended to have a higher level of profitability. We also found a significant and *negative* association between the number of favorites and the return on assets; this effect, however, disappeared after we excluded outliers in the data. A company's tweet was retweeted significantly more times during a profitable quarter than during a non-profitable quarter.

The number of likes that a company received on Facebook predicts its profitability. Specifically, we found that the average number of likes a company received each quarter on Facebook was significantly and *positively* associated with its total revenue, net income (or loss), and return on assets. Likewise, the average number of comments a company generated each quarter on Facebook was significantly and *positively* related to its quarterly total revenue; the quarterly number of shares had a significant *negative* effect on the quarterly total revenue; and average comment valence was found to be significantly but *negatively* associated with the quarterly total revenue and profit margin. The last two findings need to be interpreted with caution, because these effects disappeared when we excluded a few outliers in the data analyses, which suggested that they might be artificial. Finally, a company tended to generate more likes on Facebook during a profitable quarter, and comments on its posts during a profitable quarter were significantly more *positive* than those during a nonprofitable quarter.

The number of quarterly video likes a company received on YouTube was found to have significant and *positive* effects on its quarterly total revenue, net income (or loss), and earnings per share. The number of quarterly video dislikes on YouTube also had significant and *positive* effects on its quarterly total revenue and net income (or loss). The number of YouTube video shares had significant but *negative* effects on a company's quarterly total revenue and net income (or loss). YouTube videos in profitable quarters generated significantly more views, likes, dislikes, and comments; viewers' comments were also more *positive*.

In terms of Google+, a company's quarterly number of plusses was found to have a significant and *positive* impact on its quarterly earnings per share, and quarterly number of shares had significant and *positive* effects on its quarterly net income (or loss) and earnings per share. We found that a company's quarterly number of Google+ comments had significant but *negative* effects on its quarterly net income (or loss), earnings per share, profit margin, and return on assets.

In general, while monthly stock prices were not impacted by three of the four social media platforms, other financial variables were. Each platform significantly impacted accounting processes, and each operated differently. Several of these differences, especially those which were obtained as significantly negative, disappeared when outliers were removed from the analyses, so care should be taken when interpreting them.

Model Testing

The individual use of social media by platform clearly suggests that the two-way symmetrical model of public relations whereby stakeholders and companies interact has significant impacts on a company's financial and accounting processes. However, we were interested in testing Stacks's (2002, 2010) model of return on expectations (ROE) as a mediator on a company's return on investment (ROI) and how social media across platforms impacted on that model. The results were supportive, indicating that the original model demonstrated an impact on the company's financial outcomes, and indicated areas that may need refinement. The following discussion summarizes the findings when analyzed via the model.

As with the individual social media platform impacts on stock return, the data failed to support the hypothesized effect on the first company financial indicator for the Fortune 500 companies database. A company's nonfinancial outcomes as operationalized as the KPIs combined for the four social media platforms including reputation, relationship, trust, credibility, and confidence had no significant effect on its monthly stock return. We were disappointed in the model's lack of support for impact on stock return—only Facebook provided evidence of impact on this variable in the individual platform analyses. It may be that Facebook, as the oldest of the four platforms examined, had a historical advantage over the other three. Twitter should have enjoyed a similar advantage (it was used by 3% more companies), but it might not have been a *general* stakeholder platform; it might be more invested in the business-to-business (B2B) than the business-to-consumer (B2C) side of the business.

When examining the accounting process data, however, a number of significant *positive* findings were obtained. As with the individual platforms, ROE is related to the more specific accounting outcomes that may ultimately yield stock return, but not consistently across the model. Four of the five nonfinancial social media variables were supported in the analyses: credibility, relationship, reputation, and trust. Specifically, we found that a company's credibility, reputation and trust on social media had a significant and positive effect on its net income (or loss) and profit margin. The company's established relationship with stakeholders and trust had a significant and positive effect on its earnings per share. Additionally, a company's credibility on social media had a significant and positive effect on its total revenue, and return on assets. The public's confidence in a company on social media, however, had no significant effect on its bottom-line business performance.

There existed significant differences among the 10 business sectors regarding their social media outcomes. The consumer discretionary sector was the most successful in generating reputation, relationship, and trust on social media. Three sectors, including information technology, consumer discretionary, and consumer staples were deemed more credible on social media than others. Additionally, a company's nonfinancial activeness on four social media platforms including Twitter, Facebook, YouTube, and Google+ was significantly and positively correlated. When a company was active on one platform, it tended to be active on the other three platforms as well.

To sum up, the data analyzed in this project supported the basic ROE model and failed to support the addition of confidence to the model. The results were supportive of the nonfinancial activities associated with credibility, relationship, reputation, and trust on the accounting outcomes. However, the results did not support nonfinancial indicators on companies' stock return or performance. The only platform to obtain significant impact was Facebook and that may be a function of several factors, including historical relevance to target audience, being specialized by definition to B2C stakeholders, and/or simply a limitation placed on us by our data collection.

Research Limitations

All research has certain limitations that lead to caveats on interpreting its findings. Our research has several. First, this is not a controlled experiment. Nothing was manipulated and the study is more of a field experiment where the data grounded the interpretation(s) of social media impact. Second, given the data as obtained, commentary valence was skewed positively; therefore, one of the model components, confidence, was not clearly or cleanly operationalized. That credibility impacted on accounting processes but confidence did not support such an interpretation. Third, the database we created was huge, but perhaps neither large enough nor random enough to clearly reflect the population of social media messages. This may sound strange, but *given the negative relationships that disappeared after outliers were eliminated* from the analyses suggest the database may not have included a true reflection of the KPIs. On the other hand it may be that those outliers in their negativity clearly demonstrated the power of those comments and should be examined in more detail. And, finally, as noted in the model itself, the nonfinancial indicators of credibility, relationship, reputation, and trust are dependent on each other for outcome—that is they are intercorrelated and some results may be the impact of collinearity in the database.

Summary

Given the relative absence of research looking at the impact of top Fortune 500 companies' use of social media, the findings presented in this book should be taken as initial support for both the importance of social media in company profits and, in the case of Facebook, on stock return. The findings also support the model of public relations/corporate communication ROE on business outcomes for company accounting that may or may not mediate stock return and profits. Clearly, however, social media are an important player in corporate communication, at least among the Fortune 500 companies. In the next and last chapter, we will draw conclusions on our research and point out future research directions.

Chapter Ten

Conclusions

Chapter Overview

The impact of social media on corporate communication cannot be denied. Basic research such as this project suggests several implications for companies who wish to establish relationships with stakeholders. In this closing chapter, we look at the implications of social media communication and then make suggestions for future research. Specifically, we aim to address the following questions:

- What are the overarching implications of our research findings?
- Why does social media's influence on business performance appear to be more indirect and subtle than originally thought?
- What directions can future research look into?

Research Implications

The fact that stakeholders have a channel through the social media to freely comment on, recommend, and influence the corporation suggests that corporate communication must move toward earned and owned media and away from paid media. This is not to say that marketing products and services through advertising is dead, just that such one-way or asymmetric communication must be rethought. While it is difficult to measure social media, it is even harder to measure the impact of traditional marketing approaches. Clearly, word of mouth—and electronic word of mouth—has moved to the fore in establishing authentic and responsible companies. This is the world of corporate communication that seeks to establish authentic companies whose character influences financial outcomes as much as the product or service itself.

The findings of this project suggest, however, that the impact of social media is not as clear-cut as we thought. First, although social media are employed by the Fortune 500 companies, they do not directly affect financial outcomes. This study identified only Facebook as impacting on the metric most stakeholders consider most important: stock return. This may be because of several factors:

- The impact of Facebook being the social media platform with the longest history;
- The impact of Facebook's huge user base;
- A direct B2C orientation that precludes the impact of B2B or other forms of business;
- The ability stakeholders now have to establish deep two-way symmetrical connections with companies through Facebook.

Additionally, the fact that we looked at only Fortune 500 companies may have influenced the outcome. It may be that these top companies are less aggressive than up-and-coming companies in their use of social media.

Second, social media's influence may be more subtle than originally thought. When we looked at the impact social media have on less well-understood accounting processes—processes that from a business perspective traditionally underly the attractiveness of a company to stockholders and which are not perceived as direct but indirect influences on stock return or price—all four social media platforms studied in this project had some influence on accounting processes. And, not all platforms produced positive relationships to the financial outcomes when approached from an accounting perspective. The problem with this finding, however, was that the effect failed to be influential once outliers were removed from the analysis. This, we think, may be a function of two things: First, the sampling within the social media platforms was limited to two comments and simple counts; second, the sample itself—consisting of top companies—may represent companies that are seen by most stakeholders as practicing above average corporate communication.

Third, although the model proposed earlier in this volume was supported, there was no *direct* support for return on expectations' effect on stock return. The model suggested that confidence as a nonfinancial indicator would be reflected in a company's social media KPIs. That it was not seen as playing a significant part in the equation may be a function of the same causes listed above. However, it also may be influenced by its calculation (average of top two comments' valence scores). Given that most comments across social media platforms were neutral to positive, the effect was not as strong as it could be. Again, using a population of Fortune 500 companies may have been a factor in its failure to impact on expectations. However, the core social media nonfinancial indicators (credibility, relationship, reputation, and trust) did impact on accounting processes, and may be truly indirect and subtle influences on stock return and profit.

Suggestions for Future Research

As we have suggested throughout this volume, this project was designed as a field experiment that tested the basic premises that social media have an impact on company financial outcomes. As noted, too, we suggested several caveats when interpreting findings from the database we created. First, we looked "only" at the top Fortune 500 companies.[1] Clearly, the top companies may have an aura of their own and may also put more money, time, and thought into their social media promotional mix. A random sample of publicly traded companies may be a better base from which to study social media platform impact on business outcomes. Or, perhaps taking the Fortune 1000 companies would produce the range of companies that might yield negative comments. Also difficult to interpret is the fact that only one platform (YouTube) provides a negative indicator (closing out commentary). If all four platforms had "dislikes," "minuses," and so forth, that might lead to different findings.

Second, although the database with which we worked was extremely large, the actual numbers of comments, videos, and posts sampled was limited by our ability to code two-way symmetrical communication indicators. There was a trade-off between getting the data coded as quickly as possible, since the businesses continued to communicate during the sampling period, and having a more in-depth look at the KPIs. Future research might look at automating content coding (with sufficient comparison to human coding) to expand the valence data. It may be a good idea to redefine the valence coding from categorical (positive, negative, neutral) to an interval metric ranging from positive to negative via semantic differential-like analyses. There may be subdimensions of the KPIs that need to be explored and platforms that need to be treated as multidimensional rather than unidimensional, as we did.

Third, there may be conceptual differences between the KPIs on each social media platform, with some being rudimentary and others being more complex. For example, it requires greater mental effort on the part of the user to "comment" on a Facebook post than to "like" it.[2] As argued in many classic persuasion models, people are less willing to put effort into a task that they don't "care" about.[3] Therefore, the chances are that most comments that can be observed on the social media platforms are made by just one group of individuals—highly motivated and involved users. Interestingly, in a recent article published in *Science* (Ruths & Pfeffer, 2014), the authors pointed out that it might be problematic to use the social media population as a proxy for the real population to study human behavior, because they don't necessarily match. Thus, an examination of how companies engage stakeholders through social media that looks at their online interactions (e.g., comments and replies) may essentially be a study of how companies engage "some stakeholders" through social media. We suggest that future research take such issues into consideration and test alternative methods to avoid this kind of sampling bias.[4]

Finally, this project used many traditional statistical analysis procedures such as correlations, t-tests and ANOVA tests, and multiple regressions. They well served the purpose of analyzing the data and providing answers to our research questions. It is worth pointing out, though, that in these analyses the financial outcomes were treated as separate functions independent of each other. Perhaps more advanced multivariate analysis would be educational. If, as we suggest, the social media serve as direct and indirect influences on the business outcomes, then a structural equation model (SEM) might throw light on the findings.

Summary

This book addresses a question that to our knowledge has not been investigated with the same depth and breadth of our study. It found significant relationships between nonfinancial outcome indicators and financial outcome indicators in a social media context. The four social media platforms studied are those that are most likely to be associated with a company's social media strategies. The research findings both answered our questions and raised new questions. The test of the model of nonfinancial indicators return on expectations as influencers of business outcomes was largely supported, and suggests that future research should make use of it when examining corporate communication. Finally, this study suggests new directions for social media communication strategies. We leave it up to the readers to extend the findings of our basic research, further test the variables, and implement new strategies that will produce a better understanding of how public relations/corporate communication nonfinancial indicator variables, combined with financial indicator variables, drive business success or failure.

Notes

Chapter 1

1. As argued in Jennings, Blount, and Weatherly (2014), establishing a clear policy for employees with regard to their social media usage is vital for today's businesses because employees' activities on social media may bring unforeseen challenges to the business, ranging from financial to legal and ethical.
2. As seen in Fisher (2009), measuring ROI in social media is very important but quite difficult. Many "ROI calculators" have been developed in the past, but they are quickly dismissed and considered unworkable.
3. There has been a large amount of research on social media in the past 2 decades (see, e.g., Khang, Ki, & Ye, 2012, and Wilson, Gosling, & Graham, 2012), most of which has focused on Facebook and Twitter. In this book we included only a few of the references that are more recent and more closely relevant to our specific research.
4. How these five nonfinancial indicators are specifically defined and calculated will be explained in more detail in Chapter 8.
5. The pilot test was conducted in March 2013. The nonfinancial indicators from the five social media platforms were collected first. Then, each public company's stock price and earnings per share were recorded on March 22, 2013, and its net income (or loss) in 2012 was also recorded. Each company's ranking was based on the 2013 Fortune 500 list.

6. The pilot test suggested several significant results. First, a company's Fortune 500 ranking was significantly correlated with its numbers of Facebook likes, Twitter followers and tweets, YouTube subscribers, and Google+ plusses. Second, the company's stock price was significantly related to the numbers of Twitter followers, YouTube subscribers and video views, and Google+ plusses. Third, the net income (or loss) was significantly associated with the numbers of Facebook likes, Twitter followers and tweets, YouTube subscribers and video views, and Google+ plusses. Finally, no significant relationship was detected between a company's activities on Pinterest and its Fortune 500 ranking, stock price, earnings per share, and net income (or loss).
7. The classification of business sectors in this book was based on the Global Industry Classification Standard. The Fortune 500 companies presented in Table 1.3 include only those whose business sector coding can be accessed via Compustat, a database we used to obtain companies' financial data.

Chapter 2

1. In this study we will consider *stakeholder* and *stockholder* as synonymous. Stockholders are also stakeholders, but with a financial incentive to follow a company.
2. The effects of eWOM are well documented in the literature (e.g., Clemons, Gao, & Hitt, 2006; Duan, Gu, & Whinston, 2008; Gauri, Bhatnagar, & Rao, 2008; Li & Wang, 2013; Liu, 2006; Ye, Law, & Gu, 2009), suggesting that both valence and volume of eWOM can affect viewers' perceptions, attitudes, and behaviors.
3. The first CCO was Arthur W. Page, who served as Chief Communication Officer for AT&T from 1927 to 1947 and lends his name to the Arthur W. Page Society, which is comprised of top communication officers across a variety of industries.
4. Samples of their work and other Commission members and partners are available at the Institute's website: www.instituteforpr.org/ipr-measurement-commission.
5. Besides the public relations domain, there are also wide discussions on social media measurement in other disciplines such as advertising and marketing (e.g., Murdough, 2009; Peters, Chen, Kaplan, Ognibeni, & Pauwels, 2013; Rappaport, 2014).
6. See Stacks and Bowen, 2014.
7. The total revenue, net income (or loss), and earnings per share can be directly recorded from a publicly available database, but stock return (whether weekly, monthly, quarterly, or yearly), profit margin, return on equity, and return on assets need to be calculated.
8. As argued by Rappaport (2014), one of the major problems in social media measurement is that it is often unclear why certain metrics are selected for an examination and how they relate to an objective. To help conquer this problem, a framework should be imposed on the measurement. Consistent with this logic, we adopted

Stacks's (2002, 2011) models as the theoretical framework in this book, and the social media metrics that we examined were operationalized based on the five outcome variables depicted in Stacks (2002, 2011): credibility, reputation, trust, relationship, and confidence.

Chapter 3

1. In general, companies with established social media plans see them impact across a broad social use, while companies just starting their social media use tend to look narrowly as related to return on investment and reported as bottom line instead of triple bottom line outcome.
2. The business sector coding of each public company was recorded from the Compustat database.
3. The library at our university subscribed to the Wharton Research Data Services (http://wrds-web.wharton.upenn.edu/wrds/), which we used to access both CRSP and Compustat.
4. Both simple agreement and Cohen Kappa were used to determine coder reliability.
5. Facebook does not show a number of "followers" or "subscribers" as Twitter, YouTube, and Google+ do. Therefore, calculating the confidence variable involved KPIs of only Twitter, YouTube, and Google+.
6. Based on our data, the adoption rates of Twitter, Facebook, YouTube, and Google+ by the Fortune 500 companies was 84.0%, 81.4%, 78.2%, and 73.0%, respectively.
7. A multiple regression is a statistical procedure used to detect relationships between a dependent variable and multiple independent variables. The function of such a statistical treatment will be explained in more detail in Chapters 4 to 8.

Chapter 4

1. We used IBM-SPSS v. 22 to set up our databases and analyze the data in this book.
2. The correlation between a company's Fortune 500 ranking and its total number of tweets was found to be significant ($r = -.11, p < .05$). The correlation coefficient here was negative because a lower number represented a higher ranking.
3. A company's total number of tweets was significantly correlated with its total numbers of following ($r = .32, p < .001$) and followers ($r = .15, p < .01$). The correlation between the total numbers of following and followers was also significant ($r = .26, p < .001$).
4. A tweet's total number of replies was significantly correlated with its total numbers of retweets ($r = .28, p < .001$) and favorites ($r = .32, p < .001$). The numbers of retweets and favorites were also highly correlated ($r = .85, p < .001$).

5. To ensure that the two coders did not reach a high level of agreement by chance, we also calculated Cohen's Kappa ($k = .73$), which suggested a satisfactory level of reliability.
6. See de la Merced (2013).
7. The "Holding Period Return" was calculated based on the following formula: r(t) = [(p(t)f(t)+d(t))/p(t')] - 1, where
   r(t) = return on purchase at time t', sale at time t (t' < t)
   p(t) = last sale price or closing bid/ask average at time t
   f(t) = price adjustment factor for t
   d(t) = cash adjustment for t
   p(t') = last sale price or closing bid/ask average at time t'
8. We also recorded the monthly return of the Dow Jones Industrial Average and the Nasdaq Composite Index from January 2009 to December 2013 for the purpose of statistical control. Their effects on the Fortune 500 companies' monthly stock returns were found to be very similar to those of the Standard & Poor's 500 Composite Index in our statistical analyses. Thus, we did not report the effects of those two indexes in this book.
9. The "Return on S&P Composite Index" was calculated based on the following formula: r(t) = [SPINDX(t)/SPINDX(t')] - 1, where
   r(t) = return of the Standard & Poor's 500 Composite Index at time t (t' < t)
   SPINDX(t) = the level of the Standard & Poor's 500 Composite Index at time t
   SPINDX(t') = the level of the Standard & Poor's 500 Composite Index at time t'
10. The basic formula of a regression is y = a + bx + e, where
    y = criterion variable ("result")
    a = constant
    b = regression coefficient
    x = predictor variable ("reason")
    e = residue
    If b appears to be statistically significant, that suggests that the change of x can cause the change of y. However, we need to point out that although we considered a company's monthly stock return as y and its Twitter activeness as x, it was a statistical treatment and thus could not lead to a firm causal relationship conclusion.
11. A correlation analysis showed that a company's monthly stock return was significantly and positively correlated with the Standard & Poor's 500 Composite Index's monthly return ($r = .50, p < .001$).
12. The correlations among each tweet's numbers of replies, retweets, and favorites were presented in note 4. In addition, each tweet's average reply valence was significantly correlated with its number of favorites ($r = -.08, p < .05$), but there was no significant correlation between each tweet's average reply valence and its numbers of replies and retweets.

13. A collinearity problem is likely to occur when two or more predictor variables in a multiple regression are highly correlated together, making it difficult to draw a conclusion on what specific predictor variable holds a "true" impact on the criterion variable. After we standardized the variables in our regression analyses, we did not observe any significant collinearity problem.
14. The monthly return of the Standard & Poor's 500 Composite Index showed a significant and positive impact on the monthly stock return of the companies ($B = .46$, $\beta = .30$, $t = 9.78$, $p < .001$).
15. The p-value associated with b_2, b_3, b_4, and b_5 was .42, .74, .64, and .95, respectively.
16. A categorical predictor variable can be converted to one or more dummy variables in a regression analysis for the purpose of statistical control so that its impact on the criterion variable is accounted for. Suppose that this categorical predictor variables has n categories ($n \geq 2$), the number of dummy variables needed will be $n - 1$.
17. The number of retweets showed a significant and positive effect on the net income ($B = .06$, $\beta = .17$, $t = 2.21$, $p < .05$), profit margin ($B = .06$, $\beta = .17$, $t = 2.19$, $p < .05$), and return on assets ($B = .12$, $\beta = .25$, $t = 3.24$, $p < .01$).
18. The number of favorites showed a significant but negative impact on the return on assets ($B = -.10$, $\beta = -.21$, $t = -2.63$, $p < .01$).
19. The p-values associated with the regression coefficients of each tweet's number of replies and its overall reply valence were all above .05, suggesting statistical insignificance.
20. When there is a group of numerical cases, the average and the standard deviation of this group can be calculated. The standard deviation is an overall measure of how much distance each case is from the average. If there is a case in the group that is out of the range of the average plus or minus three times of the standard deviation, that case can be considered as an outlier. An outlier may significantly change the average of the group.
21. The primary purpose of an independent t-test is to compare two groups of cases on a certain measurable outcome.
22. There were three cases in our database where a company reported a break-even for a quarter (the net income was equal to zero). We excluded these three cases from our t-test analyses.
23. The number of retweets appeared to be significantly higher in profitable quarters ($M = 28.09$) than in nonprofitable ones ($M = 11.63$), $t(800) = 2.47$, $p < .05$ (two-tailed).
24. Profitable quarters seemed to have a higher number of replies ($M = 1.75$) and favorites ($M = 11.95$), and average reply valence ($M = .37$) than nonprofitable quarters (replies: $M = 1.27$; favorites: $M = 9.77$; average reply valence: $M = .14$). However, none of these differences was statistically significant (all p-values were above .05).

25. Different from an independent t-test, a one-way ANOVA test is to compare more than two groups of cases on a certain measurable outcome.
26. The average quarterly number of replies for each business sector appeared to be (in order): telecommunication services ($M = 10.24$), consumer staples ($M = 2.62$), consumer discretionary ($M = 2.22$), industrials ($M = 1.76$), information technology ($M = 1.09$), financials ($M = .71$), energy ($M = .59$), utilities ($M = .40$), health care ($M = .31$), and materials ($M = .28$).
27. We used the Tukey's post hoc analysis method in all the one-way ANOVA tests that we performed, so as to detect any significant difference among the 10 business sectors in regard to the number of replies, the number of retweets, the number of favorites, and the average reply valence. A Tukey's post hoc analysis suggested that the average number of replies generated by the telecommunication services sector was significantly higher than any other sector's (all p-values $< .05$).
28. The average quarterly number of retweets for each business sector appeared to be (in order): consumer discretionary ($M = 69.11$), information technology ($M = 56.26$), consumer staples ($M = 47.32$), telecommunication services ($M = 38.40$), energy ($M = 9.77$), industrials ($M = 8.31$), financials ($M = 7.79$), health care ($M = 3.63$), utilities ($M = 3.06$), and materials ($M = 2.23$). A Tukey's post hoc analysis showed that the average number of retweets generated by the consumer discretionary sector was significantly higher than all other sectors except for information technology, consumer staples, and telecommunication services (all p-values $< .05$).
29. The average quarterly number of favorites for each business sector appeared to be (in order): consumer discretionary ($M = 38.67$), information technology ($M = 26.66$), consumer staples ($M = 12.87$), telecommunication services ($M = 9.72$), industrials ($M = 3.22$), financials ($M = 1.72$), energy ($M = 1.68$), health care ($M = .76$), materials ($M = .60$), and utilities ($M = .54$). A Tukey's post hoc analysis revealed that the average number of favorites associated with the consumer discretionary sector was significantly higher than all other sectors except for information technology and telecommunication services (all p-values $< .05$).
30. The average quarterly number of reply valence for each business sector appeared to be (in order): consumer staples ($M = .35$), consumer discretionary ($M = .28$), industrials ($M = .28$), materials ($M = .24$), health care ($M = .17$), information technology ($M = .10$), financials ($M = .05$), utilities ($M = -.06$), energy ($M = -.11$), and telecommunication services ($M = -.28$). A Tukey's post hoc analysis showed that the average reply valence of consumer staples and consumer discretionary was significantly more positive than that of utilities, energy, and telecommunication services (all p-values $< .05$).
31. See Murphy (2012).

Chapter 5

1. We used the date of its very first Facebook post to determine what year a company started to use Facebook to communicate with its publics.
2. This Facebook usage history variable had eight levels in our database, with each level specifying the year a company started to use Facebook: 1 = 2014 ($N = 3$), 2 = 2013 ($N = 22$), 3 = 2012 ($N = 39$), 4 = 2011 ($N = 66$), 5 = 2010 ($N = 86$), 6 = 2009 ($N = 134$), 7 = 2008 ($N = 45$), and 8 = 2007 ($N = 12$).
3. The raw data collection from Facebook was very time-consuming, taking approximately 8 months. The total number of likes that each company generated on Facebook was recorded on a different day during this data collection period. It is worth noting that the number of likes tends to change on Facebook from time to time. Thus, the popularity ranking that we presented in this book might be different if the data were collected at a different time.
4. A company's total number of likes was found to be significantly and positively correlated with its Facebook usage history ($r = .19, p < .001$).
5. The number of likes was significantly and positively correlated with the number of comments ($r = .42, p < .001$) and the number of shares ($r = .79, p < .001$). The number of comments and the number of shares were also significantly and positively correlated with each other ($r = .41, p < .001$).
6. We also calculated Cohen's Kappa ($k = .77$) to ensure that the two coders did not reach a high level of agreement due to random chances.
7. As seen in note 5, the numbers of likes, comments, and shares were all significantly and positively correlated. In addition, the average comment valence was found to be negatively correlated with the number of comments ($r = -.03, p < .01$), but it was not significantly correlated with the numbers of likes and shares. Finally, as stated in Chapter 4, a company's monthly stock return was significantly and positively correlated with the Standard & Poor's 500 Composite Index's monthly return ($r = .50, p < .001$).
8. The monthly return of the Standard & Poor's 500 Composite Index showed a significant and positive impact on the monthly stock returns of the companies ($B = .48, \beta = .50, t = 65.83, p < .001$).
9. The hierarchical regression analysis results revealed that the number of comments had a significant and positive effect on a company's monthly stock return, and such an effect was incremental to the impact of the Standard and Poor's 500 Composite Index's monthly return ($B = .03, \beta = .03, t = 3.05, p < .01$).
10. The p-value associated with b_2, b_4, and b_5 in the hierarchical regression analysis was .44, .95, and .09, respectively, suggesting that the numbers of likes and shares and the overall comment valence had no significant effect on a company's monthly stock return.

11. The nine dummy variables for business sector and the three dummy variables for seasonality in this Facebook project were the same as those in the Twitter project, reported in Chapter 4.
12. A company's quarterly number of likes was significantly correlated with its quarterly numbers of comments ($r = .44$, $p < .001$) and shares ($r = .86$, $p < .001$.), and average comment valence ($r = .04$, $p < .01$.). Moreover, the quarterly number of comments was significantly correlated with the quarterly number of shares ($r = .42$, $p < .001$).
13. The quarterly number of likes showed significant effects on the quarterly total revenue ($B = .13$, $\beta = .13$, $t = 4.64$, $p < .001$), net income ($B = .07$, $\beta = .06$, $t = 2.07$, $p < .05$), and return on assets ($B = .06$, $\beta = .06$, $t = 2.11$, $p < .05$).
14. The quarterly number of comments had a significant effect on the quarterly total revenue ($B = .06$, $\beta = .06$, $t = 3.34$, $p < .001$).
15. The quarterly number of shares had a significant but negative effect on the quarterly total revenue ($B = -.10$, $\beta = -.10$, $t = -3.64$, $p < .001$).
16. The quarterly overall comment valence had significant but negative impacts on the quarterly total revenue ($B = -.05$, $\beta = -.05$, $t = -2.89$, $p < .01$) and profit margin ($B = -.04$, $\beta = -.04$, $t = -2.55$, $p < .05$).
17. After we deleted the outliers in the database and reconducted the hierarchical regression analyses, we found that the effect of the number of shares on the total revenue became insignificant ($p = .09$). The effects of the overall comment valence on the total revenue and profit margin also became insignificant ($p = .38$ and $p = .28$, respectively).
18. A company's quarterly number of likes in a profitable quarter ($M = 3{,}195.60$) was found to be significantly higher than that in a nonprofitable quarter ($M = 1{,}223.04$), $t(4750) = 1.97$, $p < .05$ (two-tailed).
19. The average comment valence in a profitable quarter ($M = .38$) was found to be significantly more positive than that in a nonprofitable quarter ($M = .18$), $t(4750) = 5.23$, $p < .001$ (two-tailed).
20. The number of comments in profitable quarters ($M = 274.30$) was close to that in nonprofitable quarters ($M = 276.50$), and the difference was insignificant ($p = .98$). Profitable quarters seemed to generate more shares ($M = 261.49$) than nonprofitable quarters ($M = 152.37$), but again, the difference was insignificant ($p = .37$).
21. The average quarterly number of likes for each business sector appeared to be (in order): consumer discretionary ($M = 6{,}800.28$), information technology ($M = 4{,}469.61$), telecommunication services ($M = 2{,}800.30$), consumer staples ($M = 2{,}796.04$), industrials ($M = 1{,}339.76$), financials ($M = 602.60$), energy ($M = 357.76$), health care ($M = 85.22$), materials ($M = 71.13$), and utilities ($M = 63.85$). A Tukey's post hoc analysis showed that the average number of likes generated by the consumer discretionary sector was significantly higher than all other sectors except for information technology and telecommunication services (all p-values $< .05$).

22. The average quarterly number of comments for each business sector appeared to be (in order): information technology (M = 624.63), consumer discretionary (M = 481.80), consumer staples (M = 304.99), telecommunication services (M = 279.08), industrials (M = 102.60), financials (M = 57.64), energy (M = 37.54), materials (M = 26.81), health care (M = 7.08), and utilities (M = 6.89). A Tukey's post hoc analysis showed that the average number of comments generated by the information technology sector was significantly higher than all other sectors except for consumer discretionary and telecommunication services (all p-values $< .05$).
23. The average quarterly number of shares for each business sector appeared to be (in order): information technology (M = 517.42), consumer discretionary (M = 491.98), consumer staples (M = 268.31), telecommunication services (M = 173.16), industrials (M = 76.80), financials (M = 69.51), energy (M = 26.94), health care (M = 9.44), materials (M = 9.16), and utilities (M = 6.35). The only significant differences among the 10 sectors suggested by a Tukey's post hoc analysis were that the number of shares generated by the industrial sector was significantly lower than those generated by the information technology sector ($p < .05$) and the consumer discretionary sector ($p < .05$).
24. The average quarterly comment valence for each business sector appeared to be (in order): consumer staples (M = .47), consumer discretionary (M = .46), industrials (M = .43), information technology (M = .43), energy (M = .35), health care (M = .34), materials (M = .29), financials (M = .14), utilities (M = .13), and telecommunication services (M = –.33). A Tukey's post hoc analysis showed that the average comment valence generated by the telecommunication services sector was significantly lower than all other sectors (all p-values $< .05$).

Chapter 6

1. The Fortune 500 companies adopted YouTube for their corporate communication in different years from 2005 to 2013: 2005 (N = 25), 2006 (N = 94), 2007 (N = 19), 2008 (N = 39), 2009 (N = 44), 2010 (N = 37), 2011 (N = 27), 2012 (N = 23), and 2013 (N = 13).
2. This YouTube usage history variable had nine levels in our database, each representing a different year when a company established its account on YouTube: 1 = 2013, 2 = 2012, 3 = 2011, 4 = 2010, 5 = 2009, 6 = 2008, 7 = 2007, 8 = 2006, and 9 = 2005.
3. A company's total numbers of videos, subscribers, and views on YouTube change from time to time. We recorded those three metrics for each Fortune 500 company on the YouTube website on a specific day during our data collection period. The rankings we presented in this book were based on our data. The readers should be aware that if the data were collected at different times, those rankings might be different.

4. A company's YouTube usage history was significantly correlated with its total number of subscribers ($r = .14, p < .05$).
5. A company's Fortune 500 ranking was significantly correlated with its YouTube usage history ($r = -.30, p < .001$) and total number of subscribers ($r = -.13, p < .05$). Both correlations were negative, because a lower number on the Fortune 500 list meant a higher ranking.
6. A company's total number of YouTube videos was significantly correlated with its total numbers of subscribers ($r = .28, p < .001$) and views ($r = .79, p < .001$). The total numbers of subscribers and views were also significantly correlated ($r = .28$, $p < .001$).
7. There are two possible reasons why a YouTube video has no comment: (1) no viewer is willing to comment on it, or (2) the company has disabled the comment function for this video. Therefore, we recorded whether the company had disabled the comment function for each sampled video in our database.
8. The definition of each category was the same as that used in the Twitter project and Facebook project.
9. Two coders each coded 30% of the content, and the other two coders each coded 20% of the content.
10. We compared the two types of videos (videos that allowed viewers' comments and videos that did not allow viewers' comments) in terms of the numbers of likes, dislikes, and shares. No significant difference was found with regard to the numbers of likes and shares (both p-values > .05). However, the number of dislikes when a video's comment function was disabled ($M = 20.90$) was significantly higher than when it was not ($M = 6.67$), $t(7497) = 2.70, p < .01$.
11. A YouTube video's numbers of views, likes, dislikes, comments, and shares were all significantly and positively correlated (r ranged from .54 to .92, and all p-values were <.001). In addition, a video's length was positively correlated with its number of likes ($r = .03, p < .05$), while its average comment valence was negatively correlated with its number of dislikes ($r = -.04, p < .05$). Finally, as stated in Chapter 4, a company's monthly stock return was significantly and positively correlated with the Standard & Poor's 500 Composite Index's monthly return ($r = .50, p < .001$).
12. The monthly return of the Standard & Poor's 500 Composite Index showed a significant and positive impact on a company's monthly stock return ($B = .44, \beta = .47$, $t = 32.92, p < .001$).
13. The p-values associated with b_2, b_3, b_4, b_5, b_6, b_7, and b_8 in the hierarchical regression analysis were all above .05, suggesting that the numbers of views, likes, dislikes, comments, shares, the video length, and the overall video comment valence had no significant effect on a company's monthly stock return.
14. A company's quarterly numbers of YouTube video views, likes, dislikes, comments, and shares were all significantly and positively correlated with each other (r ranged from .58 to .82, all p-values < .001). Moreover, the quarterly video length

was positively correlated with the quarterly numbers of both likes ($r = .05, p < .01$) and dislikes ($r = .04, p < .05$). Finally, the average quarterly video comment valence was positively correlated with the quarterly numbers of video views ($r = .04$, $p < .05$) and comments ($r = .04, p < .05$).

15. The nine dummy variables for business sector and the three dummy variables for seasonality in this YouTube project were the same as those in the Twitter project reported in Chapter 4.
16. The quarterly number of YouTube video likes showed significant effects on the quarterly total revenue ($B = .12$, $\beta = .11$, $t = 2.50$, $p < .05$), net income (or loss) ($B = .15$, $\beta = .10$, $t = 2.26$, $p < .05$), and earnings per share ($B = .05$, $\beta = .10$, $t = 2.19$, $p < .05$).
17. The quarterly number of YouTube video dislikes showed significant effects on the quarterly total revenue ($B = .10$, $\beta = .09$, $t = 2.87$, $p < .01$) and net income (or loss) ($B = .15$, $\beta = .10$, $t = 3.00$, $p < .01$).
18. The quarterly number of YouTube video shares showed significant but negative effects on the quarterly total revenue ($B = -.11$, $\beta = -.11$, $t = -3.45$, $p < .001$) and net income (or loss) ($B = -.11$, $\beta = -.08$, $t = -2.46$, $p < .05$).
19. None of the YouTube activeness indicators showed a significant impact on the quarterly profit margin, return on assets, or return on equity (all p-values > .05).
20. To test whether the significant effects of the number of YouTube video dislikes on the total revenue and net income (or loss) were caused by a few outliers in the database, we reconducted the regression analyses excluding those outliers. However, the findings remained the same (p-values associated with both effects < .001).
21. As argued by Berger, Sorensen, and Rasmussen (2010), negative publicity can potentially increase consumers' purchase likelihood by increasing product awareness.
22. To test whether the significant effects of the number of YouTube video shares on the total revenue and net income (or loss) were caused by a few outliers in the database, we reconducted the regression analyses excluding those outliers. The effect on the net income (or loss) became insignificant ($p = .55$), but the effect on the total revenue was still significant ($p < .001$).
23. Profitable quarters generated significantly more video views ($M = 83{,}219.76$), video likes ($M = 195.41$), video dislikes ($M = 19.59$), and video comments ($M = 30.62$) than nonprofitable quarters (video views: $M = 22{,}421.09$; video likes: $M = 50.80$; video dislikes: $M = 5.41$; video comments: $M = 15.80$) (all p-values < .05, two-tailed). In addition, the video comments were significantly more positive in profitable quarters ($M = .17$) than in nonprofitable quarters ($M = .12$) ($p < .05$, one-tailed). Finally, there was no significant difference between the two types of quarters in regard to the number of video shares and the video length (both p-values > .05).
24. The average quarterly YouTube video length (counted in seconds) for each business sector appeared to be (in order): information technology ($M = 641.31$),

health care (M = 454.70), materials (M = 448.53), industrials (M = 443.22), utilities (M = 411.83), financials (M = 408.61), energy (M = 374.32), consumer staples (M = 309.86), telecommunication services (M = 301.10), and consumer discretionary (M = 296.75). A Tukey's post hoc analysis showed that the videos produced by the information technology sector were significantly longer than the videos produced by any other sector (all p-values < .05).

25. The average quarterly number of YouTube video likes for each business sector appeared to be (in order): information technology (M = 772.81), consumer discretionary (M = 159.69), industrials (M = 98.16), financials (M = 39.11), consumer staples (M = 34.14), telecommunication services (M = 22.31), energy (M = 20.85), materials (M = 10.89), health care (M = 6.03), and utilities (M = 4.61). A Tukey's post hoc test revealed that the number of video likes generated by the information technology sector was significantly higher than all other sectors (all p-values < .05).
26. The average quarterly number of YouTube dislikes for each business sector appeared to be (in order): information technology (M = 79.72), consumer discretionary (M = 13.69), energy (M = 8.90), telecommunication services (M = 6.16), industrials (M = 6.14), consumer staples (M = 4.29), financials (M = 4.21), materials (M = 3.98), utilities (M = .79), and health care (M = .66). A Tukey's post hoc test revealed that the number of video dislikes generated by the information technology sector was significantly higher than all other sectors (all p-values < .05).
27. The average quarterly number of YouTube video views for each business sector appeared to be (in order): information technology (M = 248,756.95), consumer discretionary (M = 81,782.41), financials (M = 58,400.26), energy (M = 45,827.75), consumer staples (M = 44,240.28), industrials (M = 41,830.43), materials (M = 14,887.59), telecommunication services (M = 9,342.65), health care (M = 7,607.77), and utilities (M = 3,199.26). A Tukey's post hoc test showed that the number of video views generated by the information technology sector was significantly higher than all other sectors (all p-values < .05), except for energy and telecommunication services.
28. The average quarterly number of YouTube video comments for each business sector appeared to be (in order): information technology (M = 114.85), consumer discretionary (M = 26.77), industrials (M = 21.86), telecommunication services (M = 10.74), consumer staples (M = 7.28), financials (M = 4.88), energy (M = 4.32), materials (M = 1.52), utilities (M = 1.44), and health care (M = 1.10). A Tukey's post hoc test suggested that the number of video comments generated by the information technology sector was significantly higher than all other sectors (all p-values < .01), except for telecommunication services.
29. The average quarterly YouTube video comment valence for each business sector appeared to be (in order): consumer discretionary (M = .31), industrials (M = .26), consumer staples (M = .19), information technology (M = .14), energy (M = .12), financials (M = .10), materials (M = .07), health care (M = .06), utilities (M = .03),

and telecommunication services ($M = -.01$). A Tukey's post hoc test showed that the comments on the videos produced by the consumer discretionary sector were significantly more positive than the comments on the videos produced by any other sector (all p-values $< .01$), except for industrials.

30. The average quarterly number of YouTube video shares for each business sector appeared to be (in order): information technology ($M = 207.06$), financials ($M = 42.23$), consumer discretionary ($M = 29.69$), consumer staples ($M = 27.42$), industrials ($M = 9.90$), health care ($M = 8.56$), energy ($M = 7.62$), telecommunication services ($M = 5.24$), materials ($M = 2.36$), and utilities ($M = .91$). The only significant difference suggested by a Tukey's post hoc analysis was that the number of video shares generated by the information technology sector was significantly higher than that of consumer discretionary and industrials (both p-values $< .05$).

Chapter 7

1. Each company's total number of Google+ followers was recorded from the Google+ website on a specific day during our data collection period. The rankings that we presented in this book were based on those data. The readers should be aware that if the data were collected at different times, the rankings might be different.
2. A company's number of followers on Google+ was significantly and positively correlated with its number of views ($r = .63$, $p < .001$) and number of people in its circles ($r = .23$, $p < .01$). Moreover, a company's number of Google+ followers was significantly and negatively correlated with its Fortune 500 ranking ($r = -.21$, $p < .001$). This correlation was negative because a lower number on the Fortune 500 list meant a higher ranking.
3. Counting from November 2011 to December 2013, there are 26 months. Thus, a maximum of 26 posts could be sampled for a company.
4. A Google+ post's number of plusses was significantly and positively correlated with its numbers of comments ($r = .69$, $p < .001$) and shares ($r = .81$, $p < .001$). The number of comments and the number of shares were also significantly correlated ($r = .70$, $p < .001$).
5. The definition of each category was the same as that used in the other three projects (Twitter, Facebook, and YouTube).
6. As seen in note 4, a post's numbers of plusses, comments, and shares were significantly and positively correlated. Moreover, a post's average comment valence was significantly and positively correlated with its numbers of plusses ($r = .12$, $p < .001$) and shares ($r = .07$, $p < .05$). Finally, as stated in Chapter 4, a company's monthly stock return was significantly and positively correlated with the Standard & Poor's 500 Composite Index's monthly return ($r = .50$, $p < .001$).

7. The monthly return of the Standard & Poor's 500 Composite Index showed a significant and positive effect on the monthly stock return of the companies ($B = .42$, $\beta = .37$, $t = 20.12$, $p < .001$).
8. The p-values associated with b_2, b_3, b_4, and b_5 in the hierarchical regression analysis were all above .05, suggesting that the numbers of plusses, comments, and shares, and the overall comment valence had no significant impact on a company's monthly stock return.
9. A company's quarterly numbers of plusses, comments, and shares were all significantly and positively correlated (r ranged from .68 to .84, all p-values $< .001$). In addition, the average quarterly comment valence was significantly and positively correlated with the quarterly numbers of plusses, comments, and shares (r ranged from .12 to .29, all p-values $< .001$).
10. The quarterly number of Google+ plusses showed a significant and positive effect on the quarterly earnings per share ($B = .03$, $\beta = .20$, $t = 2.63$, $p < .01$).
11. The quarterly number of Google+ shares showed a significant and positive effect on the quarterly net income (or loss) ($B = .14$, $\beta = .19$, $t = 2.19$, $p < .05$) and earnings per share ($B = .05$, $\beta = .36$, $t = 4.36$, $p < .001$).
12. The quarterly number of Google+ comments had a significant but negative impact on the quarterly net income (or loss) ($B = -.13$, $\beta = -.17$, $t = -2.65$, $p < .01$), earning per share ($B = -.03$, $\beta = -.23$, $t = -3.79$, $p < .001$), profit margin ($B = -.09$, $\beta = -.15$, $t = -2.26$, $p < .05$), and return on assets ($B = -.13$, $\beta = -.16$, $t = -2.43$, $p < .05$).
13. The quarterly total revenue and return on equity were not significantly predicted by any of the four Google+ activeness indicators (all p-values $> .05$).
14. After we deleted the outliers in the database and reconducted the hierarchical regression analyses, we found that the negative effects of the number of Google+ comments on the net income (or loss), profit margin, and return on assets became insignificant (all p-values $> .05$), but the effect on earnings per share remained significant ($p < .01$).
15. We conducted a correlation analysis with a company's quarterly number of Google+ comments and its six quarterly financial indicators (total revenue, net income or loss, earnings per share, profit margin, return on equity, and return on assets). The only significant correlation was between the number of comments and earnings per share ($r = .11$, $p < .01$), and the correlation was in fact positive. Thus, we suspect that the positive association between the number of comments and earnings per share became negative in the regression analyses because of a collinearity problem (a company's numbers of plusses, comments, and shares were highly correlated, and their effects on the company's earnings per share were likely to be strongly intertwined).
16. The independent t-test results suggested that a company's quarterly Google+ activeness in a profitable quarter (plusses: $M = 121.25$; comments: $M = 19.11$; shares: $M = 19.57$; and the average comment valence: $M = .29$) did not significantly

differ from that in a nonprofitable quarter (plusses: $M = 156.89$; comments: $M = 39.94$; shares: $M = 25.45$; and the average comment valence: $M = .28$) (all p-values > .05, two-tailed).

17. The average quarterly number of plusses for each business sector appeared to be (in order): information technology ($M = 212.73$), consumer discretionary ($M = 204.39$), consumer staples ($M = 144.41$), telecommunication services ($M = 56.52$), industrials ($M = 46.34$), energy ($M = 12.74$), financials ($M = 10.33$), materials ($M = 3.94$), health care ($M = 3.76$), and utilities ($M = 1.86$). The only significant differences, suggested by a Tukey's post hoc analysis, were that the numbers of plusses generated by the information technology and consumer discretionary sectors were significantly higher than those generated by the industrials and financials sectors (all p-values < .05).
18. The average quarterly number of comments for each business sector appeared to be (in order): telecommunication services ($M = 38.32$), information technology ($M = 33.46$), consumer discretionary ($M = 30.39$), consumer staples ($M = 30.23$), industrials ($M = 7.59$), financials ($M = 2.16$), energy ($M = 1.41$), materials ($M = .75$), health care ($M = .48$), and utilities ($M = .25$). A Tukey's post hoc analysis showed that the number of comments generated by the information technology sector was significantly higher than those generated by the industrials and financials sectors (both p-values < .05). Also, the number of comments generated by the consumer discretionary sector was significantly higher than that generated by financials ($p < .05$).
19. The average quarterly number of shares for each business sector appeared to be (in order): information technology ($M = 45.95$), consumer discretionary ($M = 24.24$), consumer staples ($M = 18.00$), industrials ($M = 6.28$), telecommunication services ($M = 5.60$), energy ($M = 3.07$), financials ($M = 1.59$), materials ($M = 1.56$), utilities ($M = .88$), and health care ($M = .64$). The only significant differences detected in a Tukey's post hoc analysis were that the number of shares generated by the information technology sector was significantly higher than those generated by the industrials and financials sectors (both p-values < .05).
20. The average quarterly comment valence for each business sector appeared to be (in order): consumer discretionary ($M = .52$), industrials ($M = .47$), consumer staples ($M = .45$), information technology ($M = .22$), energy ($M = .15$), materials ($M = .06$), utilities ($M = .00$), financials ($M = -.01$), health care ($M = -.02$), and telecommunication services ($M = -.20$). A Tukey's post hoc analysis revealed that the comments on the posts by the consumer discretionary, industrials, and consumer staples sectors were significantly more positive than those on the posts by the financials, health care, and telecommunication services sectors (all p-values < .05).
21. See Miller (2011).
22. As suggested by Cohen (1992), when the significance criterion (e.g., $\alpha = .05$) and power (e.g., power of .80) are set, a smaller effect will need a larger sample to

detect. In other words, if the sample size is not large enough, it may not catch a small but significant effect.

Chapter 8

1. *Operationalization* is a term frequently used in social science research. It represents the process of giving a specific definition to a broad and abstract concept (which is typically called a "construct"), and this definition must accurately reflect the fundamental meaning of the concept.
2. The correlations among relationship, reputation, trust, credibility, and confidence were all significant (r ranged from .21 to .78, all p-values $< .01$), except for the correlation between trust and confidence ($p = .46$).
3. A company's Fortune 500 ranking was significantly correlated with its reputation, trust, and credibility (r ranged from –.13 to –.10, all p-values $< .001$). These correlations were negative because a lower number on the Fortune 500 list meant a higher ranking. No significant correlations were found between a company's Fortune 500 ranking and its relationship and confidence (both p-values $> .05$).
4. The monthly return of the Standard & Poor's 500 Composite Index had a significant and positive impact on a company's monthly stock return in all five sets of hierarchical regressions (all p-values $< .001$).
5. None of the outcome measures, including relationship, reputation, trust, credibility, and confidence, had a significant effect on a company's monthly stock return (all p-values $>$ 05).
6. A company's quarterly reputation had significant effects on its quarterly net income (or loss) ($B = .13$, $\beta = .18$, $t = 2.58$, $p < .05$) and profit margin ($B = .11$, $\beta = .16$, $t = 2.21$, $p < .05$).
7. A company's quarterly relationship had a significant effect on its quarterly earnings per share ($B = .03$, $\beta = .21$, $t = 3.16$, $p < .01$).
8. A company's quarterly trust had significant effects on its quarterly net income (or loss) ($B = .14$, $\beta = .19$, $t = 2.92$, $p < .01$), earnings per share ($B = .03$, $\beta = .21$, $t = 3.05$, $p < .01$), and profit margin ($B = .11$, $\beta = .15$, $t = 2.16$, $p < .05$).
9. A company's quarterly credibility had significant influences on its quarterly total revenue ($B = .11$, $\beta = .10$, $t = 7.11$, $p < .001$), net income (or loss) ($B = .11$, $\beta = .09$, $t = 6.35$, $p < .001$), profit margin ($B = .09$, $\beta = .09$, $t = 5.84$, $p < .001$), and return on assets ($B = .10$, $\beta = .09$, $t = 6.07$, $p < .001$).
10. A company's quarterly confidence had no significant impacts on any of its business performance measures, including the quarterly total revenue, net income (or loss), earnings per share, profit margin, return on equity, and return on assets (all p-values $> .05$).

11. We compared different business sectors on their standardized outcome scores by conducting five sets of one-way ANOVA tests. Based on the ANOVA tests and post hoc analyses, significant differences were detected for relationship, reputation, trust, and credibility, but no significant difference was found for confidence. The significant results are summarized as follows: (a) the relationship score of consumer discretionary was significantly higher than those of industrials and financials (both p-values $< .05$); (b) the reputation score of consumer discretionary was significantly higher than those of all other sectors except for materials and telecommunication services (all p-values $< .05$); (c) the trust score of consumer discretionary was significantly higher than those of industrials, financials, and information technology (all p-values $< .05$); (d) the credibility scores of information technology, consumer discretionary, and consumer staples were significantly higher than those of energy, utility, materials, industrials, health care, and financials (all p-values $< .001$).
12. The numbers of Twitter replies, Facebook comments, YouTube comments, and Google+ comments were all significantly and positively correlated (r ranged from .05 to .52, all p-values $< .05$).
13. The numbers of Twitter favorites, Facebook likes, YouTube likes, and Google+ plusses were all significantly and positively correlated (r ranged from .06 to .57, all p-values $< .01$), with the exception of the number of Facebook likes and the number of YouTube likes ($p = .09$).
14. The numbers of Twitter retweets, Facebook shares, YouTube shares, and Google+ shares were all significantly and positively correlated (r ranged from .17 to .30, all p-values $< .001$), except that the number of YouTube shares was not significantly correlated with the number of Twitter retweets ($p = .24$) and the number of Facebook shares ($p = .76$).
15. A company's numbers of Twitter followers, YouTube subscribers, and Google+ followers were significantly and positively correlated (r ranged from .27 to .46, all p-values $< .001$).
16. We conducted correlation analyses with the average valence scores of Twitter replies, Facebook comments, YouTube comments, and Google+ comments. It was found that the average valence score of Facebook comments was significantly correlated with that of Twitter replies ($r = .10$, $p < .01$) and that of YouTube comments ($r = .07$, $p < .01$). No other significant correlation was detected (all p-values $> .05$).
17. The chi-square test is often used to test the differences across several groups (in our case, it referred to different business sectors) on an outcome that is measured in a categorical manner (in our case, it referred to a business sector's adoption rate of social media).
18. The chi-square test result suggested that there were significant differences across various business sectors regarding their social media adoption rates ($\chi^2 = 122.43$, $df = 36$, $p < .001$).

19. According to the one-way ANOVA test result, no significant association was found between a company's social media adoption rate and its Fortune 500 ranking ($p = .79$). The average Fortune 500 ranking at five social media adoption levels was similar: (a) no social media presence (average ranking = 273.42), (b) presence on one social media platform (average ranking = 257.86), (3) presence on two social media platforms (average ranking = 265.12), (4) presence on three social media platforms (average ranking = 243.22), and (5) presence on all four social media platforms (average ranking = 247.40).
20. The numbers of Twitter replies, Twitter favorites, and Twitter retweets were all significantly correlated with the variable "time" (r ranged from –.17 to –.06, all p-values < .001). The correlations were negative because a smaller value in "time" represented a more recent month.
21. The numbers of Facebook comments, Facebook likes, and Facebook shares were all significantly with the variable "time" (r ranged from –.19 to –.05, all p-values < .05). The correlations were negative because a smaller value in "time" represented a more recent month.
22. The number of YouTube likes and the number of YouTube dislikes were found to be significantly correlated with the variable "time" (both $r = -.03$, both p-values < .05). Moreover, the YouTube video length was significantly correlated with the variable "time" ($r = -.03$, $p < .01$). The correlations were negative because a smaller value in "time" represented a more recent month.
23. The number of Google+ comments was significantly and positively correlated with the variable "time" ($r = .06$, $p < .01$), whereas the number of Google+ plusses was significantly and negatively correlated with the variable "time" ($r = -.04$, $p < .05$). Since a smaller value in "time" represented a more recent month, these correlations meant that as time went by, the public tended to give fewer comments but more plusses to a company's Google+ posts.

Chapter 9

1. As opposed to advertising, which is a one-way communication strategy whereby the message is carefully created and placed in the media, public relations and corporate communication are both "owned" and "earned," while advertising messages are "paid."
2. For an excellent overview and cases of situations where social media communication resulted in individual and company crises, see Capozzi and Rucci (2013).
3. As seen in Pornpitakpan (2004), source credibility is a central concept examined in persuasive communication research. It has two major dimensions: expertise and trustworthiness. A "third-party endorser" is generally regarded as credible, likely because he or she is trustworthy (not because of his or her expertise).

Chapter 10

1. Although our research has certain limitations, it is also true that this research is much bigger in scope than most prior research found in the literature, so we believe that the findings presented in this book make a unique contribution to the field. Taking two previous studies in public relations and marketing as examples, Rybalko and Seltzer (2010) examined how the Fortune 500 companies communicated with their publics by sampling 930 tweets from 93 companies, and Swani, Milne, and Brown (2013) discussed the Fortune 500 companies' message strategies on Facebook via a content analysis of 1,146 Facebook posts from 280 companies.
2. According to Gerlitz and Helmond (2013), the "like" button on Facebook was introduced as a "shortcut" to commenting. This function may be more attractive to social media "lurkers" because they are less likely to provide in-depth comments.
3. According to the Elaboration Likelihood Model (ELM) (Petty & Cacioppo, 1986), a well-established persuasion framework, people are likely to be involved in more elaborative information processing (the central route) when they are highly involved with the subject. On the other hand, when they are not involved with the subject, their information processing tends to be superficial (the peripheral route) and they use various "cues" to make quick judgments and decisions.
4. As argued by Ruths and Pfeffer (2014), researchers need to think of ways to reduce biases and flaws in collecting and evaluating social media data. One of the suggested approaches is to test robustness of research findings across different platforms and different time frames. We consider our book to be an initial test of the relationship between corporate social media use and business bottom-line measures. We encourage other scholars to replicate our research using similar methods but incorporating different enterprises (e.g., randomly selected public corporations), different social media platforms (e.g., LinkedIn, Instagram), and different time spans (e.g., from 2014).

References

Alexa. (2014). *The top 500 sites on the Web*. Retrieved from http://www.alexa.com/topsites

Altimeter. (2013). *The state of social business 2013: The maturing of social media into social business* [Report]. Retrieved from http://www.slideshare.net/Altimeter/report-the-state-of-social-business-2013-the-maturing-of-social-media-into-social-business

Barwise, P., & Meehan, S. (2010). One thing you must get right when building a brand. *Harvard Business Review, 88*(12), 80–84.

Berger, J., Sorensen, A. T., & Rasmussen, S. J. (2010). Positive effects of negative publicity: When negative reviews increase sales. *Marketing Science, 29*(5), 815–827.

Berthon, P. R., Pitt, L. F., Plangger, K., & Shapiro, D. (2012). Marketing meets Web 2.0, social media, and creative consumers: Implications for international marketing strategy. *Business Horizons, 55*(3), 261–271.

Bond, R. M., Fariss, C. J., Jones, J. J., Kramer, A. D. I., Marlow, C., Settle, J. E., & Fowler, J. H. (2012). A 61-million-person experiment in social influence and political mobilization. *Nature, 489*(7415), 295–298.

Boyd, D. M., & Ellison, N. B. (2007). Social network sites: Definition, history, and scholarship. *Journal of Computer-Mediated Communication, 13*(1), 210–230.

Capozzi, L., & Rucci, S. (2013). *Crisis management in the age of social media*. New York: Business Expert Press.

Chaney, P. (2009). *The digital handshake: Seven proven strategies to grow your business using social media*. Hoboken, NJ: Wiley.

Chen, Z., & Lurie, N. H. (2013). Temporal contiguity and negativity bias in the impact of online word of mouth. *Journal of Marketing Research, 50*(4), 463–476.

Cho, J., Park, D. J., & Ordonez, Z. (2013). Communication-oriented person-organization fit as a key factor of job-seeking behaviors: Millennials' social media use and attitudes toward organizational social media policies. *Cyberpsychology, Behavior and Social Networking, 16*(11), 794–799.

Christ, P. (2005). Internet technologies and trends transforming public relations. *Journal of Website Promotion, 1*(4), 3–14.

Clemons, E. K., Gao, G. G., & Hitt, L. M. (2006). When online reviews meet hyperdifferentiation: A study of the craft beer industry. *Journal of Management Information Systems, 23*(2), 149–171.

Cohen, J. (1992). A power primer. *Psychological Bulletin, 112*(1), 155–159.

Corcoran, S. (2009). Defining earned, owned and paid media [Web log post]. Retrieved from http://blogs.forrester.com/interactive_marketing/2009/12/defining-earned-owned-and-paid-media.html

Cutlip, S. M. (1994). *The unseen power: Public relations: A history*. Hillsdale, NJ: Lawrence Erlbaum Associates.

Davidson, W. P. (1983). The third-person effect in communication. *Public Opinion Quarterly, 47*(1), 1–15.

de la Merced, M. J. (2013, October 29). Sale of Dell closes, moving company into private ownership. *The New York Times.* Retrieved from http://dealbook.nytimes.com/2013/10/29/sale-of-dell-closes-moving-company-into-private-ownership/

DiStaso, M., & McCorkindale, T. (2013). A benchmark analysis of the strategic use of social media for Fortune's most admired US companies on Facebook, Twitter and YouTube. *Public Relations Journal, 7*(1), 1–33.

Duan, W., Gu, B., & Whinston, A. B. (2008). Do online reviews matter? An empirical investigation of panel data. *Decision Support Systems, 45*(4), 1007–1016.

Duncan, T., & Caywood, C. (1996). The concept, process and evolution of integrated marketing communication. In E. Thorson & J. Moore (Eds.), *Integrated communication: Synergy of persuasive voices* (pp. 13–34). Mahwah, NJ: Lawrence Erlbaum Associates.

Dutta, S. (2010). What's your personal social media strategy? *Harvard Business Review, 88*(11), 127–130.

Edosomwan, S., Prakasan, S. K., Kouame, D., Watson, J., & Seymour, T. (2011). The history of social media and its impact on business. *Journal of Applied Management and Entrepreneurship, 16*(3), 79–89.

Eisend, M. (2010). Explaining the joint effect of source credibility and negativity of information in two-sided messages. *Psychology & Marketing, 27*(11), 1032–1049.

Fisher, T. (2009). ROI in social media: A look at the arguments. *Journal of Database Marketing & Customer Strategy Management, 16*(3), 189–195.

Gauri, D. K., Bhatnagar, A., & Rao, R. (2008). Role of word of mouth in online store loyalty: Comparing online store ratings with other e-store loyalty factors. *Communications of the ACM, 51*(3), 89–91.

Gerlitz, C., & Helmond, A. (2013). The like economy: Social buttons and the data-intensive web. *New Media and Society, 15*(8), 1348–1365.

Global Industry Classification Standard. (2014). *GICS structure*. Retrieved from http://www.msci.com/products/indexes/sector/gics/gics_structure.html

Goodman, A. (2013, September 25). *Top 40 Buffett-isms: Inspiration to become a better investor*. Retrieved from http://www.forbes.com/sites/agoodman/2013/09/25/the-top-40-buffettisms-inspiration-to-become-a-better-investor/2/

Grunig, J. (1984). Organizations, environments, and models of public relations. *Public Relations Research and Education, 1*(1), 6–29.

Grunig, J., & Grunig, L. (1992). Models of public relations and communications. In J. E. Grunig (Ed.), *Excellence in public relations and communication management* (pp. 285–326). Hillsdale, NJ: Lawrence Erlbaum Associates.

Grunig, L. A., Grunig, J., & Dozier, D. (2002). *Excellent public relations and effective organizations: A study of communication management in three countries*. Mahwah, NJ: Lawrence Erlbaum Associates.

Hansson, L., Wrangmo, A., & Søilen, K. S. (2013). Optimal ways for companies to use Facebook as a marketing channel. *Journal of Information, Communication and Ethics in Society, 11*(2), 112–126.

Hayes, R. (2014). *Corporate share values to achieve stakeholder engagement*. Presentation at the 2nd International Public Relations Summit, Yogyakarta, Indonesia.

Hoffman, D. L., & Fodor, M. (2010). Can you measure the ROI of your social media marketing? *MIT Sloan Management Review, 52*(1), 41–49.

Jennings, S. E., Blount, J. R., & Weatherly, M. G. (2014). Social media—a virtual Pandora's box: Prevalence, possible legal liabilities, and policies. *Business and Professional Communication Quarterly, 77*(1), 96–113.

Kaplan, A. M., & Haenlein, M. (2010). Users of the world, unite! The challenges and opportunities of social media. *Business Horizons, 53*(1), 59–68.

Khang, H., Ki, E., & Ye, L. (2012). Social media research in advertising, communication, marketing, and public relations, 1997–2010. *Journalism & Mass Communication Quarterly, 89*(2), 279–298.

Kiron, D., Palmer, D., Phillips, A. N., & Berkman, R. (2013). *Social business: Shifting out of first gear*. MIT Sloan Management Review research report. Boston: Massachusetts Institute of Technology.

Li, C., & Wang, X. (2013). The power of eWOM: A re-examination of online student evaluations of their professors. *Computers in Human Behavior, 29*(4), 1350–1357.

Lillqvist, E., & Louhiala-Salminen, L. (2014). Facing Facebook: Impression management strategies in company-consumer interactions. *Journal of Business and Technical Communication, 28*(1), 3–30.

Lin, K. Y., & Lu, H. P. (2011). Intention to continue using Facebook fan pages from the perspective of social capital theory. *Cyberpsycholoy, Behavior, and Social Networking, 14*(10), 565–570.

Liu, Y. (2006). Word of mouth for movies: Its dynamics and impact on box office revenue. *Journal of Marketing, 70*(3), 74–89.

Mamic, L. I., & Almaraz, I. A. (2013). How the larger corporations engage with stakeholders through Twitter. *International Journal of Market Research, 55*(6), 851–872.

McCombs, M., & Shaw, D. L. (1972). The agenda-setting function of mass media. *Public Opinion Quarterly, 36*(2), 176–187.

McKinsey & Company. (2013). Bullish on digital: McKinsey global survey results. Retrieved from http://www.slideshare.net/GaldeMerkline/mckinsey-bullish-on-digital

Michaelson, D., & Stacks, D. W. (2014). *A professional and practitioners guide to public relations research, measurement, and evaluation* (2nd ed.). New York: Business Expert Press.

Miller, C. C. (2011, June 28). Another try by Google to take on Facebook. *The New York Times*. Retrieved from http://www.nytimes.com/2011/06/29/technology/29google.html?pagewanted=all

Murdough, C. (2009). Social media measurement: It's not impossible. *Journal of Interactive Advertising, 19*(1), 94–99.

Murphy, D. (2012, September 22). Twitter building Tweet-downloading archive feature by year's end. *PC Magazine*. Retrieved from http://www.pcmag.com/article2/0,2817,2410062,00.asp

Nelson-Field, K., Riebe, E., & Sharp, B. (2012). What's not to "like?": Can a Facebook fan base give a brand the advertising reach it needs? *Journal of Advertising Research, 52*(2), 262–269.

Neuwirth, K., & Frederick, E. (2002). Extending the framework of third-, first-, and second-person effects. *Mass Communication and Society, 5*(2), 113–140.

Nugroho, F. (2014, November 4). *Social media measurement in economic terms*. Presentation at the 2nd International Public Relations Summit, Yogyakarta, Indonesia.

Paek, H. J., Hove, T., Jung, Y., & Cole, R. T. (2013). Engagement across three social media platforms: An exploratory study of a cause-related PR campaign. *Public Relations Review, 39*(5), 526–533.

Peters, K., Chen, Y., Kaplan, A. M., Ognibeni, B., & Pauwels, K. (2013). Social media metrics—A framework and guidelines for managing social media. *Journal of Interactive Marketing, 27*(4), 281–298.

Petty, R. E., & Cacioppo, J. T. (1986). *Communication and persuasion: Central and peripheral routes to attitude change*. New York: Springer-Verlag.

Pornpitakpan, C. (2004). The persuasiveness of source credibility: A critical review of five decades' evidence. *Journal of Applied Social Psychology, 34*(2), 243–281.

Raacke, J., & Bonds-Raacke, J. (2008). MySpace and Facebook: Applying the uses and gratifications theory to exploring friend-networking sites. *Cyberpsychology and Behavior, 11*(2), 169–174.

Rappaport, S. D. (2014). Lessons learned from 197 metrics, 150 studies, and 12 essays: A field guide to digital metrics. *Journal of Advertising Research, 54*(1), 110–118.

Rogers, E. (2003). *The diffusion of innovations* (5th ed.). New York: Simon & Schuster.

Ruths, D., & Pfeffer, J. (2014). Social media for large studies of behavior. *Science, 346*(6213), 1063–1064.

Rybalko, S., & Seltzer, T. (2010). Dialogic communication in 140 characters or less: How Fortune 500 companies engage stakeholders using Twitter. *Public Relations Review, 36*(4), 336–341.

Saxton, G. D., & Waters, R. D. (2014). What do stakeholders like on Facebook? Examining public reactions to nonprofit organizations' informational, promotional, and community-building messages. *Journal of Public Relations Research, 26*(3), 280–299.

Schultz, D. E., & Schultz, H. F. (1998). Transitioning marketing communication into the twenty-first century. *Journal of Marketing Communications 4*(1), 9–26.

Smith, A. N., Fischer, E., & Chen, Y. (2012). How does brand-related user-generated content differ across YouTube, Facebook, and Twitter? *Journal of Interactive Marketing, 26*(2), 102–113.

Stacks, D. W. (2002). *Primer of public relations research*. New York: Guilford.

Stacks, D. W. (2011). *Primer of public relations research* (2nd ed.). New York: Guilford.

Stacks, D. W., & Bowen, S. A. (2014). *Dictionary of public relations measurement and research* (3rd ed.). Retrieved from http://www.instituteforpr.org/dictionary-public-relations-measurement-research-third-edition/

Stacks, D. W., & Michaelson, D. (2010). *A practitioner's guide to public relations research, measurement, and evaluation*. New York: Business Expert Press.

Swani, K., Milne, G., & Brown, B. P. (2013). Spreading the word through likes on Facebook: Evaluating the message strategy effectiveness of Fortune 500 companies. *Journal of Research in Interactive Marketing, 7*(4), 269–294.

Toubia, O., & Stephen, A. T. (2013). Intrinsic vs. image-related utility in social media: Why do people contribute content to Twitter? *Marketing Science, 32*(3), 368–392.

Tuten, T. L., & Solomon, M. R. (2013). *Social media marketing*. Upper Saddle River, NJ: Pearson Education.

VisionCritical. (2013). *From social to sale: 8 questions to ask your customers*. Retrieved from http://www.visioncritical.com/sites/default/files/pdf/whitepaper-social-to-sale.pdf

Wallace, E., Buil, I., de Chernatony, L., & Hogan, M. (2014). Who "likes" you and why? A typology of Facebook fans from "fan"-atics and self-expressives to utilitarians and authentics. *Journal of Advertising Research, 54*(1), 92–109.

Waters, R. D., Burnett, E., Lamm, A., & Lucas, J. (2009). Engaging stakeholders through social networking: How nonprofit organizations are using Facebook. *Public Relations Review, 35*(2), 102–106.

Weiner, M. (2006). *Unleashing the power of PR: A contrarian's guide to marketing and communication*. San Francisco: International Association of Business Communicators.

Wilson, R. E., Gosling, S. D., & Graham, L. T. (2012). A review of Facebook research in the social sciences. *Perspectives on Psychological Science, 7*(3), 203–220.

Ye, Q., Law, R., & Gu, B. (2009). The impact of online user reviews on hotel room sales. *International Journal of Hospitality Management, 28*(1), 180–182.

Index

1

A

B

C

D

E

F

G

H

I

J

K

L

M

N

O

P

Q

R

S

T

U

V

W

Y